The Institute of Biology
Studies in Biology

Mari...pl

John H. Wickstead
D.Sc., Ph.D., F.I.Biol.

Marine Biological Association of the
United Kingdom

© John H. Wickstead 1976

First published 1976
by Edward Arnold (Publishers) Limited
25 Hill Street, London W1X 8LL

Boards edition ISBN: 0 7131 2548 9
Paper edition: ISBN: 0 7131 2549 7

Printed in Great Britain by
The Camelot Press Ltd, Southampton

General Preface to the Series

It is no longer possible for one textbook to cover the whole field of Biology and to remain sufficiently up to date. At the same time teachers and students at school, college or university need to keep abreast of recent trends and know where the most significant developments are taking place.

To meet the need for this progressive approach the Institute of Biology has for some years sponsored this series of booklets dealing with subjects specially selected by a panel of editors. The enthusiastic acceptance of the series by teachers and students at school, college and university shows the usefulness of the books in providing a clear and up-to-date coverage of topics, particularly in areas of research and changing views.

Among features of the series are the attention given to methods, the inclusion of a selected list of books for further reading and, wherever possible, suggestions for practical work.

Readers' comments will be welcomed by the author or the Education Officer of the Institute.

1976

The Institute of Biology,
41 Queens Gate,
London, SW7 5HU.

Preface

The zooplankton is of fundamental importance to the proper understanding of the seas. For the taxonomist virtually every phylum is represented. For the ecologist various communities and habitats exist. For the population dynamicist much is to be learnt from the fluctuations of the many species present. For the fisheries biologist the zooplankton forms a vital link between the primary producers and the fish. The artist will never cease to be amazed at the almost infinite variation in beauty, line and form. One could add even more to the list. However, from whatever aspect the zooplankton is viewed, it is a great asset to have some knowledge of it as an entity before the more specialized eye is turned on it. This book attempts to provide this.

Plymouth,
1976

J. H. W.

Contents

1 History of Zooplankton Research

1.1 Origins

The sea has always had an influence upon the development of man, yet until recently man's knowledge of the sea, including his knowledge of plankton, was very limited indeed. The very presence of plankton was hardly suspected until about 150 years ago, and certainly its importance was not recognized until about the turn of the century. Various phenomena associated probably with plankton had been noted by mariners during historic times; thus Pytheas, in about the fourth century B.C., stated that the sea beyond Thule (probably in the Arctic region) became thick and sluggish, like neither land nor sea. This can be related to the account of the scientific cruise of the Italian corvette *Vettor Pisani* between 1882 and 1885. To quote from Haeckel's famous *Planktonic Studies*, during this cruise, 'Sometimes the water seems coagulated, jelly-like, even to the touch.' For a long time the blue-green alga *Oscillatoria (Trichodesmium) erythraeum*, has been known to British sailors as 'sea-sawdust', and was referred to by Captain Cook during his third voyage.

Thus, no doubt, ever since man first paddled himself along on a log, isolated planktonic phenomena have been noted by seafaring men, but they had no inkling of the extent, numbers and organization of plankton communities. Leeuwenhoeck, famous for his *Little Animals*, examined some sea water with his microscope and, according to a translation of Letter 18, '. . . discovered in it divers living animalcules'. This was, however, literally, a random dip, and most of Leeuwenhoeck's observations related to fresh water. Another Dutchman, named Slabber, published in 1778 a work entitled *Natural Amusements and Microscopical Observations*, in which was the first description and illustration of a crab zoea (Fig. 1–1); but he did not appreciate what the zoea was.

It was not until the turn of the nineteenth century that ideas on the extent of the plankton community came to be mooted. The pioneer of systematic and accurately recorded observations on plankton appears to be John V. Thompson. His work has been unknown until relatively recently, but he was the first to observe live and metamorphosing crab zoeae and megalopae, and realized that they are not related animals, but stages in the life history of the crab (Fig. 1–2). In addition he was the first to observe and understand metamorphosis in acorn barnacles (Fig. 1–3). The majority of plankton books do not mention his name but his work, undertaken during his official appointment as Deputy Inspector General

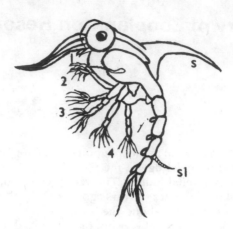

Fig. 1–1   Drawing of a crab zoea in 1778 by Slabber.

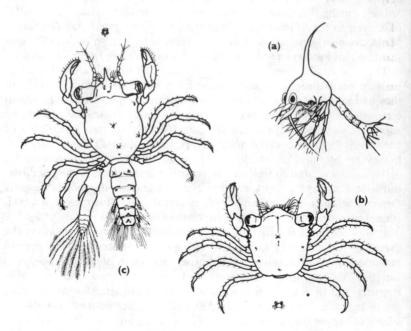

Fig. 1–2   Thompson's drawing of crab zoeae in 1835. (a) zoea, (b) megalopa, (c) young crab. The large appendage in (c) is Thompson's original Fig. 3 labelled 'one of its subabdominal fins [*sic*] more highly magnified'.

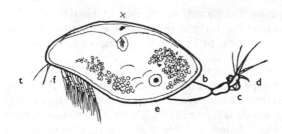

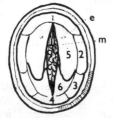
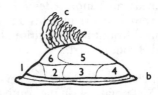

Fig. 1-3   Thompson's drawing of acorn barnacle metamorphosis, 1828-34.

of Hospitals, should be the starting-point for every plankton worker. His
Memoir 1, *On Zoea, &c.*, published in 1828, appears to be the very first
mention in the literature of a plankton tow, or 'towing' net.

Charles Darwin, during his *Beagle* voyage, made various observations
but only dabbled in plankton work. However, he did anticipate the
plankton net; his notes of December 1833 state, '. . . I often towed astern
a net made of bunting, and thus caught many curious animals'.

1.2   Period of development

The door to widespread plankton investigations was opened by
Johannes Müller of Berlin. In the autumn of 1845, at Helgoland, he
began his 'pelagic fishery by means of a fine net'. The catch was then
referred to as 'pelagic tow-stuff' (*pelagische Auftrieb*). This collecting
method soon became widespread and, to quote Haeckel, 'A new and
inexhaustibly rich field was thus opened to zoötomical and microscopical
investigation, and anatomy and physiology, organology and histology,
ontogeny and systematic zoology have been advanced to a surprising
degree'. Plankton studies progressed rapidly, but worldwide aspects of
the plankton were not appreciated fully until the *Challenger* expedition of
1873-6. Important observations were made and Murray (later Sir John)
noted, 'Everywhere we have found a rich organic life at and below the
surface of the ocean'. Some idea was now being obtained as to the extent,
in three dimensions, of the plankton, but it is interesting to note that

Alexander Agassiz maintained to the end of his days that between the plankton at or near the surface, down to say 200 fathoms, and the fauna on or near the bottom, there was a vast region where virtually no life existed.

Victor Hensen and his followers, the so-called Kiel School, made the first serious attempt to quantify the study of plankton. It was beginning to be appreciated that the zooplankton was the intermediate step in the food chain through which fish populations were able to feed from the prime producers, the phytoplankton. Hensen reasoned, by extrapolation, that if one knew the size of the plankton population, then the size of the fish population would follow. By using this argument Hensen in 1889 obtained an unprecedented sum of money from the German Government to organize his 'Plankton Expedition'. Valuable results were obtained from this Expedition, but too much was read into the results. Haeckel commented,

'(1) The most important generalizations which the plankton expedition of Kiel obtained on the composition and distribution of the plankton in the ocean stand in sharp contradiction to all previous experience; one or the other is wrong. (2) It seems to me that Hensen has incautiously founded a number of far-reaching erroneous conclusions on very insufficient premises. Finally, I am convinced that the whole method employed by Hensen for determining the plankton is utterly worthless, and that the general results obtained thereby are not only false, but also throw a very incorrect light on the most important problems of pelagic biology.'

There were many fierce, and at times bitter, arguments in the field of planktology at this time; but these did serve the purpose of being stimulating to the advance of the science.

During this period many terms that we know today came into being, generally as defined by Haeckel. The term 'plankton' was coined by Hensen in 1886 to replace 'Auftrieb', but was rather ill-defined. In 1890 Haeckel re-defined plankton, the word itself being from the Greek and meaning 'wandering' or 'roaming', and gave it its present-day meaning. It was also Haeckel who brought forward the words 'benthos' (bottom-living animals) and 'nekton' (free-swimming animals). The terms nanoplankton, microplankton, mesoplankton, macroplankton, and so on, all relate to size ranges. There are, unfortunately, no agreed limits to these dimensions. The smallest size of organisms is the nanoplankton, say up to about 25 μm. A size range should always be quoted when these terms are employed.

Realization of the scale and character of oceanographical investigations led to the Stockholm Conference of 1899 and the subsequent foundation in 1901 of the International Council for the Exploration of the Sea (I.C.E.S.), with its headquarters at Charlottenlund

Slot, near Copenhagen. A number of International Stations for continued observations were established as a result of this conference and it is interesting to note that International Station E.1, at 50° 02′N–4° 22′W, near Plymouth, is probably the most worked station in the world.

By 1899 marine laboratories had been widely established as bases from which to conduct oceanographical research. Concarneau, probably the oldest, was established in 1859, Kiel in 1868, Naples in 1872, Woods Hole, U.S.A., in 1888, and in Great Britain, Millport (or its forerunner) in 1885, Plymouth in 1888 and Aberdeen in 1899. Many more laboratories have been built in various parts of the world since the turn of the century and at the present day there are more than 2000 devoted to some aspect of marine research.

1.3 Present-day plankton research

Much of the earlier plankton work was devoted to identification and classification of the many organisms new to science; this was given impetus by the new concepts of evolution initiated by Darwin. Today, more emphasis is given to the study of plankton as a dynamic entity. No population of animals is static and that of zooplankton fluctuates a great deal, both in the short and long term. Any alteration of the factors surrounding a population will cause it to change in some way. Much research is devoted to finding out how changes in the various environmental factors affect the zooplankton and what changes can be expected as a result.

Much more thought is now being given to what the seas could yield in the way of food for an ever-increasing human population, and much effort is being devoted to resolving such problems as how much food can be taken from the seas, what sustained harvest can the seas stand, and how food production from the sea can be boosted.

Aquaculture is now being studied intensively in many parts of the world. It must be remembered when setting up schemes such as the culture of prawns, oysters, mussels, and indeed most fish, that all these animals start their lives as components of plankton; so a knowledge of zooplankton is clearly important for culture projects. Various suggestions have been put forward with the aim of greatly increasing plankton production in a given area of ocean resulting in comparably increased fish population. These schemes vary from pumping deep-oceanic nutrient-rich water up into enclosed lagoons of atolls to sinking a small nuclear pile into the depths so that the heat evolved will cause extensive upwelling.

2 The Environment of the Zooplankton

2.1 Water movements

About 70% of the earth's surface is covered with water to an average depth of about 3800 metres (12 500 feet), the volume being about 1370×10^6 cubic kilometres (325×10^6 cubic miles). As the result of various physical factors there is a basic system of water movements imprinted on the oceans. Figures 2–1 and 2–2 show the basic system of surface currents, essentially wind derived, and deep-water movements, essentially the result of different physical properties of water masses. While one talks of deep-water movements, it is perhaps more appropriate to speak of deep-water drifting. Even so, photographs of the ocean bed show, in some instances, mud ripples or bending pennatulids (sea pens), indicating a significant water flow.

Generally the surface currents correspond to prevailing winds, forming a basic pattern of nearly closed loops moving clockwise in the northern hemisphere and counter-clockwise in the southern hemisphere. The force available from the wind to drive the currents is very small, so the earth's rotation has a profound effect on these wind-driven currents. This is the so-called 'Coriolis force' or 'effect', named after the French scientist Coriolis. He published the mathematics of this geostrophic acceleration in 1835, but it was in fact a restatement of the original findings of La Place in 1775. The Coriolis force has a minimal effect at the equator, and increases with latitude. It is a deflecting force which acts on all moving masses on the earth, causing deflection to the right in the northern hemisphere and to the left in the southern hemisphere, hence the respective clockwise and counter-clockwise rotations.

Recently a new current system of great importance has been discovered and named the Cromwell Current after its discoverer. This current forms a thin ribbon of swiftly flowing water about 200 metres thick and about 300 kilometres wide, moving eastwards nearly symmetrically beneath the equator. The depth it occupies is somewhat variable, but it can be detected as little as 20 metres below the surface. The rate of flow can be up to 3 knots and it can transport some 40 000 000 cubic metres of water per second. An interesting point is that this current is being considered seriously as a future energy source. First found in the Pacific in 1951, subsequent investigations have shown similar currents to exist in the Atlantic and Indian Oceans.

Since the zooplankton depends on the current systems of the oceans for its distribution it can be seen that there is plenty of scope for this distribution.

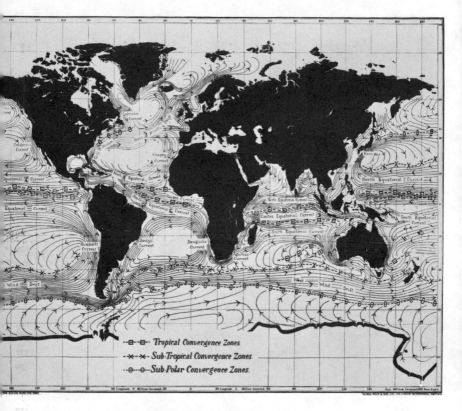

Fig. 2–1 Basic system of surface currents. (By kind permission of the Trustees of the British Museum (Natural History).)

Water movements are also extremely important to the fertility of the sea. Life in the sea depends ultimately on chlorophyll-containing plants utilizing the sun's energy for photosynthesis. Light energy is absorbed fairly rapidly by water (see p. 16) and the compensation depth is soon reached; this is the depth at which the amount of food used in the metabolism of the plant equals the amount of food made by photosynthesis. Clearly, many factors will affect the clarity of the water and the dependent compensation depth; but even in the clearest ocean water it is not much below 200 metres. Thus, effectively, all food chains in the oceans depend on the photosynthetic activities of plants within the surface 200 metres or so. There is, however, more to it than this. While light energy is one prerequisite for photosynthesis, also essential for protein synthesis are the various nutrients such as nitrates and phosphates. We thus have a situation in which the nutrients are being used above the compensation depth in the 'euphotic zone', but not below it. When this richer water penetrates into the euphotic zone, then there will be a more rapid growth of plants and a resulting larger population of

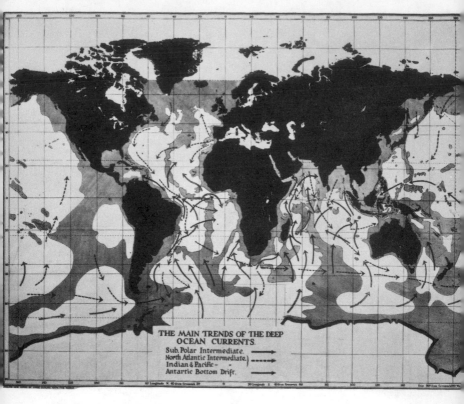

Fig. 2–2   Basic system of deep ocean current trends. (By kind permission of the
Trustees of the British Museum (Natural History).)

the animals which feed on the plants. This transport of nutrients can be
brought about by water movements in various ways, both on a large and a
small scale.

Perhaps the best known is the phenomenon of upwelling. This can
occur near the coast or in the open sea, but is most commonly associated
with coastal wind-drift currents. It is in fact a complex phenomenon, but
basically it is the situation where a steady off-shore wind causes a drift of
surface water away from the coast, and the water is then replaced by
deeper water coming to, or near to, the surface (Fig. 2–3).

Since well before World War II ideas have been put forward for
inducing significant artificial upwelling with the aim of increasing the
productivity of the seas. Feasibility studies have been undertaken for a
nuclear power station to be sited on the sea bed in about 75 metres of
water, the cooling water to be drawn in from near the bottom and
expelled towards the surface. What is perhaps the simplest means—in
theory—of inducing artificial upwelling is a phenomenon called the
perpetual salt fountain, described as an oceanographical curiosity. It

works on the principle that in many tropical and subtropical oceanic regions the salinity of the warmer surface layers exceeds that of the colder water below. If the surface water is connected to the deep water by a tube and the deep water pumped *slowly* to the surface to allow for the temperature of the water within the tube to equate with the temperature outside, and the pump is then disconnected, the water will continue to flow indefinitely. Interestingly the reverse applies, so if the surface water were pumped downwards, then this would also continue to flow. This principle can be demonstrated experimentally (Fig. 2–4). Note that some patience is required. An 8-centimetre diameter vertical glass tube is half filled with hot fresh water. The same quantity of coloured cold fresh water is introduced slowly through a hole B in the bottom. This is done carefully to avoid excessive mixing at the interface. An inner tube, C, restricted at the top to suppress secondary flows inside the tube, is lowered gently into the water, a finger being held over the upper end of the tube until the lower end is below the interface. In this way, the inner tube is filled with the cold coloured water. If this is done slowly enough for the water in the tube to equalize temperature with the water outside at the same level, the water from the tube will stay above the outside interface, even after the top of the tube is below the surface of the upper layer. After a few moments, small secondary movements and circulations are visible inside C, probably associated with the slow warming and cooling of the system as a whole. Now a salinity difference is introduced. A little hot salt water is added with gentle stirring to the upper layer outside tube C, care being taken to maintain the overall stability of the layers. Almost immediately, a stream of coloured bottom water starts to flow from the top of the tube C. This is the situation shown in the Figure. A third layer forms at the very top, this being warm, fresh and coloured, and the flow continues until the conditions of the experiment break down.

Basically, this compares with the tropical Atlantic: warm saline water above, cold, less saline water below. The cold, less saline water is drawn up the tube into the warm saline water zone. The tube permits heat, but not salt, diffusion. Hence, on attaining temperature equilibrium, the density within the tube is less than outside, and therefore rises steadily, drawing up more cold, less saline water which, in turn, equilibrates with the outside, warmer temperature, thus maintaining the flow.

If an island is situated in the path of a current, surface water accumulates on the up-current side while some upwelling of nutrient-rich water occurs on the down-current side (Fig. 2–5). In addition, the smooth flow of the current is upset, resulting in swirls and eddies which help to bring nutrient-rich water to the surface. A result of this is that the lee side of islands and promontories in the path of an established current are important areas of organic production.

There are many examples of submarine hills or sea mounts rising to within a few hundred metres of the surface, particularly in the Pacific Ocean. A number of these are very close to the equator in the path of the

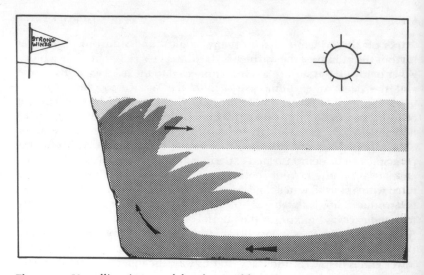

Fig. 2–3 Upwelling is caused by deep cold ocean water replacing warmer surface water driven offshore by winds. The upper layer represents the euphotic zone. (After La Fond.)

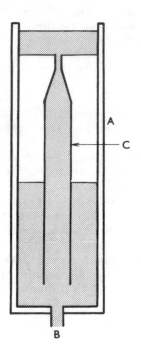

Fig. 2–4 The laboratory experiment to demonstrate the salt fountain. (After Stommel, Arons and Blanchard.)

Cromwell Current. This results in turbulence and vertical water movements which bring nutrient-rich water towards the surface over and on the lee side of the sea mounts (Fig. 2–6). It is often possible to detect the presence of a sea mount lying in the path of an established current by the aggregation of sea birds in the vicinity.

Under certain conditions convection cells can be established (Fig. 2–7). The sea surface can lose an appreciable amount of heat, particularly at night, by radiation, evaporation and conduction. The consequent increase in the density of the surface layer causes it to sink. Warmer water must rise to take its place, and thus convection cells are formed, usually some 100 to 200 metres across, the denser water sinking at the edges, and the less dense replacement water rising in the centre. The water is affected to a maximum depth of about 100 metres. The boundary of two convection cells, where the water is sinking, is often marked by a line of flotsam or a slick; and these boundaries between adjoining convection cells can be rich sources of plankton.

Thermoclines are important factors in oceanography (see p. 13). Since these are often established at shallow depths it may be wondered why convection cells cannot prevent them forming, or break them down. Basically the problem is one of energy exchange. If the sea absorbs more energy during the day than is lost at night, stable conditions are built up and a thermocline can be established. On the other hand, if the energy loss at night exceeds the energy gain during the day, then conditions will be too unstable for a thermocline to be established.

These are just some illustrations of how the seas are in a state of constant motion, both vertically and horizontally. For a fuller account of these movements, which are really very complex, reference should be made to a book on physical oceanography (see 'Further Reading'). The important point which must be remembered is that the oceans are always moving; some of these movements can be local and restricted, but others are on a very large scale.

2.2   Some physical and chemical properties

Sea water is a solution containing at least traces of nearly all the known elements. The most common ions present are chloride, sulphate, sodium, magnesium and potassium which respectively constitute 55%, 7.7%, 30.6%, 3.7% and 1.1% of the total dissolved solids, and the major constituents of sea water are present in the same proportions virtually throughout the oceans. Exceptions are where land drainage brings down a lot of land-derived salts, particularly of calcium and magnesium, to coastal areas.

Salinity is defined as the total amount, in grammes, of dissolved salts, contained in one kilogramme of sea water, i.e. as parts per thousand (°/oo). The normal range of salinities is from about 37°/oo, found for instance in the eastern Mediterranean, through 35°/oo, which is about average, to about 34°/oo in many tropical seas. Modern analytical methods, which measure the conductivity, can detect differences of 0.02°/oo, or sometimes less, which are significant in locating different water masses in deep water.

Fig. 2–5   Diagram of upwelling in the lee of an island. The upper layer represents the euphotic zone. (After La Fond.)

Routine oceanographic investigations usually include what are referred to as TS (temperature and salinity) measurements or profiles. The reversing thermometers used for temperature recording can be accurate to $\pm 0.02°C$. Sea surface temperatures vary between about 28°C

Fig. 2–6   Sea mount in path of current causing deep nutrient-rich water to rise into the surface nutrient-poor water. The upper layer represents the euphotic zone. (After La Fond.)

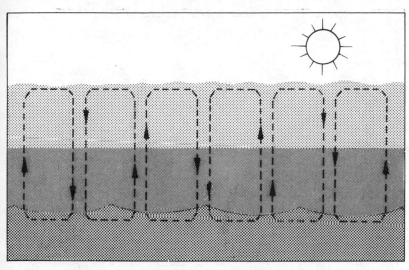

Fig. 2–7　Convection cells. The upper layer represents the euphotic zone. (After La Fond.)

in many tropical areas down to about −2°C in polar regions; deep-water temperatures can be 2°C or less.

Calculated directly from the salinity and temperature is the density. Density is defined as mass per unit volume and is usually measured in grammes per cubic centimetre: sea water has a density of about 1.02500. For convenience the whole number is ignored and the decimal point shifted three places to the right, the resulting figure being referred to as the σ_t (sigma-tee). Thus a density of 1.02714 would be 27.14 σ_t.

The sea is not an homogeneous water mass; bodies of water can be separated by differing physical and/or chemical characteristics in both a horizontal and a vertical direction. One of the most important boundaries is a thermocline, shown diagrammatically in Fig. 2–8. In this particular example the temperature drops slowly and steadily for the first 130 metres or so, below which there is a very rapid drop in the region of the thermocline itself until a depth of about 170 metres is reached, after which the temperature again decreases slowly and steadily with depth. In shallow water, a thermocline is usually seasonal and is established during periods of stable weather, being broken down eventually by strong winds. The topmost thermocline can occur from a few metres below the surface down to several hundred metres, the temperature difference being up to 15°C or more and the depth through which the temperature drop occurs being anything from one or two metres to fifty or more. One of the routine oceanographic exercises is to establish the depth of the thermocline by means of a bathythermograph, which is essentially a temperature-sensitive element connected to a moving pointer which

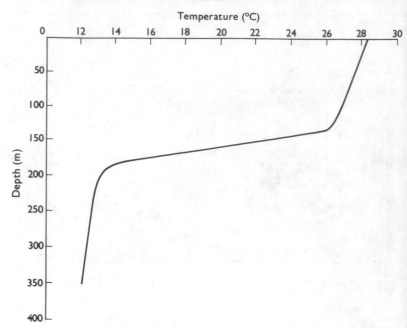

Fig. 2–8 A thermocline occurring between 130 m and 180 m in this example.

records a trace on a piece of film or smoked glass. An example of such a trace is shown in Fig. 2–9.

A thermocline can separate appreciable temperature and density differences, and thus forms an effective barrier against the vertical movement of many plankton animals.

Changes in temperature and/or salinity will also affect the viscosity. In general, tropical waters are less dense and have a lower viscosity than colder waters. This results in tropical zooplankton being smaller and/or possessing more extensions of the body or skeletal surface, thus increasing the surface area/volume ratio.

Many physical processes occur at the surface which cannot occur deeper down, e.g. heating, cooling, evaporation, and if a body of water sinks, as for instance at the Antarctic Convergence, there will be little energy available to cause this body of water to mix with others. Thus the many different bodies of water tend to retain characteristic physical features, identifiable by analysis, and characteristic plankton populations, which help to differentiate between them.

It is easy to demonstrate how different bodies of water remain discrete. Add a litre of fresh water to a large circular trough and through a tube carefully introduce a litre of a coloured solution containing 20 grammes

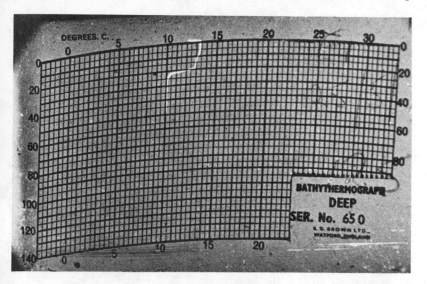

Fig. 2–9   A bathythermograph trace of a thermocline. This is a recording taken at the International Station E.1 (see p. 5).

of salt. To the bottom of this again add a litre of a different coloured solution containing 40 grammes of salt. This should stabilize to give three distinct layers; more layers could be added if necessary.

Salts dissolved in water have the effect of lowering the freezing-point. Sea water with a salinity of $32\%_0$ begins to freeze at $-1.74°C$, and with $40\%_0$ at $-2.20°C$. It is necessary to use the expression 'begins to freeze' since only the water freezes, the salts staying in solution in the remaining water. This depresses further the freezing-point of the remaining water, and so there is formed a strong brine solution which becomes more and more difficult to freeze.

All atmospheric gases can be found in solution in sea water, carbon dioxide being present in large quantities, chiefly in combination as carbonates and bicarbonates. Solution occurs at the air-sea interface, the amount dissolved being determined by the water temperature and salinity. The concentration of gases not involved in vital processes remains fairly constant, but oxygen and carbon dioxide are both very much involved in life processes and their concentrations vary accordingly. Very high concentrations of phytoplankton, which evolve oxygen as a product of photosynthesis, can cause the water to be super-saturated in oxygen, and thus able to support a larger than normal zooplankton population. Conversely, water can be de-oxygenated by decomposition of organic matter. If this occurs in more or less enclosed water, large de-oxygenated zones can result. The Black Sea is a little over

2000 metres deep and is isolated from the Mediterranean by the Bosphorus Ridge, only some 40 metres deep. This prevents any influx of oxygen-rich water into the Black Sea. The thermal conditions and the flow of fresh water into the Black Sea from rivers prevent any deep aeration through convection currents. The effect of this is that the decomposition of organic material settled from the productive upper layers has formed an anaerobic, hydrogen sulphide-rich, body of water some 1800 metres deep. Therefore animal life is confined to the top 150 to 200 metres. Similar, but not so marked, conditions can exist in some nearly-enclosed fjords.

Only a small amount of carbon dioxide remains as dissolved gas, most forming bicarbonates and carbonates. All the reactions involved are reversible, the whole forming what is known as the Carbon Dioxide System; this plays a most important part in regulating the acidity (pH) of sea water. In the upper layers pH is normally between about 8.1 and 8.3. Pressure affects the pH value by lowering it to about 7.8. Both upper and lower limits can vary slightly according to local circumstances.

2.3   Light

All the zooplankton depends ultimately on the ability of the phytoplankton to photosynthesize. Clearly, therefore, it is necessary to consider factors controlling the amount of light entering the sea.

Of the energy from the sun which reaches the sea surface some 50% is infra-red and virtually complete absorption of this occurs within the top metre. Fairly rapid absorption occurs at the ultra-violet end of the spectrum. Nearly all the radiation capable of penetrating the sea to any appreciable extent falls within the visible spectrum. As indicated below (p. 17) blue light, about 450–475 nm, penetrates the furthest (Fig. 2–10).

The plant pigment directly concerned with photosynthesis is chlorophyll. There are several forms of chlorophyll, but they all show maximum light absorption at the red end of the spectrum. Theoretically this would restrict the phytoplankton to shallower regions of the seas than they in fact are, but other pigments are present in addition to the chlorophylls. These pigments are not directly involved with photosynthesis, but are considered to be able to absorb alternative wave lengths and to transfer the energy to the chlorophylls, in this way making it available for photosynthesis. This extends the depth at which photosynthesis is possible but, as will be seen from Fig. 2–11, which shows available energy, it is still restricted to about 200 metres in ideal conditions.

The 24-hour cycle of the light penetrating the sea has a great influence on the behaviour of the zooplankton. This is discussed in greater detail in the section on vertical migration (p. 19).

Daylight has been detected down to about 1000 metres (see Fig. 2–11).

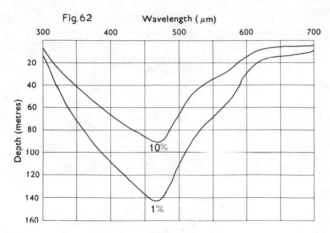

Fig. 2–10   The depths at which the percentage of surface radiation is 10% and 1% respectively for different wavelengths in clearest ocean water. (After Jerlov.)

At and below this depth bioluminescence interferes with readings of daylight penetration. The near-end region of light penetration is referred to as the twilight zone, and it is in this zone particularly that the eyes of many deep sea animals are very large relative to body size.

A proportion of the light entering the sea has been found to be polarized. Some crustaceans and cephalopods have a visual mechanism which can perceive the polarization and are reported to use this for 'navigation' and orientation.

Where some degree of light is available many animals have the capability of hunting visually, and different protective devices have been evolved in order to counter this. One of the commonest is transparency. Many of the zooplankton are transparent, some to such a degree that they are virtually impossible for at least the human eye to detect when alive and in sea water. Also various colour patterns and pigmentations can be adopted for camouflage. Surface zooplankton often have a deep blue pigment while many animals living in the blue twilight zone below about 500 metres are red since, with blue light only being present, the animals appear black, and thus are invisible. This can be demonstrated easily by directing light of about 460 nm on to a colour plate of deep sea prawns in a dark room. Many migratory animals are also red. A near-surface, night-time oceanic plankton sample is usually a basic red or orange-red. Just as some animals can match the level of light around them by means of photophores (p. 54), so can some of the nearer-surface fish, including some of the post-larval forms which can be regarded as planktonic, match the surrounding light intensity by the reflectivity of their scales. Guanine crystals, which give the silvery appearance, are so arranged that the fish

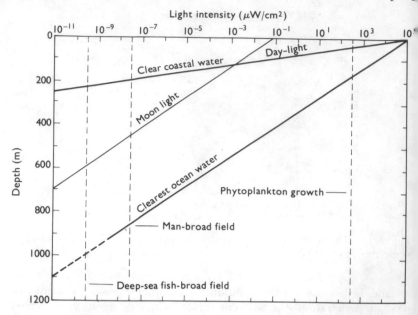

Fig. 2–11 Light energy penetration into the clearest ocean water and into clear coastal water. The graph shows the minimum underwater light intensity values necessary for vision in man and some deep-sea fish. (After Clarke and Denton.)

can be rendered invisible, or at least inconspicuous, against a light background. Alternatively, as a defence mechanism, if a predator attacks a shoal of silvery fish, the shoal disperses or changes direction rapidly to the accompaniment of many random 'flashes' of silver, which serve to confuse the attacker.

A characteristic feature of many animals living close to the surface is development of a deep blue pigment. This is considered by many to have been developed as a protection against ultraviolet penetration. Certainly it does not appear to serve as camouflage since the blue colour makes them conspicuous from under the water.

Light is clearly a vital factor for virtually all living organisms and, whether 'natural' or 'produced', as in bioluminescence, plays a major role in the *modus vivendi* of the zooplankton.

3 Zoological Composition

3.1 Temporary and permanent plankton

Numerous organisms pass their complete life-cycle in the plankton. These constitute the permanent plankton, or holoplankton. Examples are the chaetognaths, pteropods, larvaceans, siphonophores and many copepods.

Alternatively many planktonic animals pass only part of their life-cycle in the plankton, as egg, larva or adult, or as combinations of these stages. These constitute the temporary plankton, or meroplankton. Examples of these are acorn barnacles, whose larva is planktonic; many fish in which the egg and early larva are planktonic; various coelenterates which pass part of their alternation of generations as benthic animals; many decapod crustaceans, in which eggs and larvae can be planktonic, and some parasitic copepods, in which the larva and breeding adult are planktonic. Near-shore and/or shallow-water plankton samples show greater numbers and variety of meroplanktonic animals than do oceanic samples.

Another type of meroplankton can be recognized. For this type the phase spent as plankton depends on the time of the day or year. Many mysids and ostracods live on or in the bottom during the day but rise and enter the plankton community during the night. *Tretomphalus*, a warm water foraminiferan, lives normally as a benthic animal attached to a substrate, but during the breeding season, a terminal gas-filled chamber is developed and the animal releases itself from the substrate and enters the plankton. Similarly syllid and nereid worms—typical benthic animals—are well known for their incursions into the plankton during their breeding swarms.

3.2 Day and night variation

There must be very few, if any, plankton animals which are not sensitive to light in some degree. Generally they are both photopositive (move towards light) and photonegative (move away from light), the members of each species tending to move up or down to congregate at what Rose described as their 'optimum lumineux caractéristique'. Some plankton animals seek a maximum level of illumination and so are found at the surface. Others are sensitive to low, medium or high levels of illumination and/or different wavelengths. Since static physical conditions do not exist in the sea there will be constant movement of the zooplankton to maintain or attain their position at the level of optimum light intensity. As the angle of incidence of light on the water varies continuously throughout the day due to the apparent movement of the sun, there are vertical plankton movements based on the 24-hour sun cycle, but these are not simple and straightforward. Various other factors can influence the cycle, such as the boundary of a thermocline. Some factors within the 24-hour light cycle also affect the vertical distribution of the zooplankton, e.g. varying amounts of cloud cover obscuring the direct sun, the state of the sea surface which can cause variations in the degree of reflection and scatter, and therefore light penetration, and the phase of the moon.

Although the zooplankton consists of weak swimmers, the swimming performance is only weak when measured against the horizontal current systems in which they live. When related to their size many of the zooplankton are powerful swimmers and can move vertically through relatively great distances. The larger planktonic crustaceans such as euphausiids can move upwards at speeds of between 100 and 400 metres an hour. Medium sized copepods can move upwards at between 30 and 60 or more metres an hour, while even the nauplius of the acorn barnacle (*Balanus*) has been observed to move as fast as 10 to 15 metres in an hour. Aided by gravity, downward movement is even quicker.

The general picture is a general pattern of movement towards the surface during the darkness followed by a withdrawal from the surface during daylight.

It must not be assumed that the zooplankton actually reaches the surface. Different groups of animals move through different vertical distances in different sections of the water column, the overall result being what has been called 'The Ladder of Migrations' (Fig. 3–1). This vertical migration can be seen as a means whereby food resources in the surface layers of the oceans are made available to the deep-living animals.

Clearly the movement of zooplankton will affect the content of plankton samples from any given level during a 24-hour period. In shallow coastal water the permanent plankton animals will be at different levels according to the surface illumination, therefore a horizontal plankton sample taken at a fixed level throughout a 24-hour period will show varying quantities of zooplankton present at different times of the day (Fig. 3–2). If vertical samples are taken from the bottom to the top of the water column, this would catch the zooplankton at whatever level they were; but during the daytime the temporary plankton animals resting on or in the bottom would be missed. These examples give some indication of the problems of getting reliable quantitative plankton samples.

In open oceanic waters there is a marked increase in the plankton taken at night. Table 3–1 shows the yearly average of plankton from an Indian Ocean station; note the greater percentage increase in dry weight at night at both depths. This is due to relatively large animals joining the near-surface plankton.

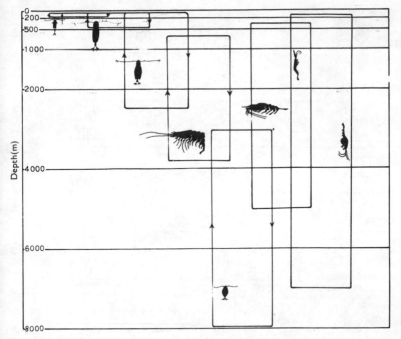

Fig. 3–1 Ladder of Migrations (after Vinogradov) showing how various planktonic forms move through different depth zones.

It is perhaps worth noting that plankton animals are sensitive to varying degrees of moonlight; a clear, full-moon night will keep migratory zooplankton well below their 'darkness level'.

Table 3–1 The differences between day and night plankton at a tropical oceanic station in the top 50 m and 200 m. Results are mean figures per m^3 for 13 months of sampling with a 70 cm 74 mesh/inch net.

	P.S 3 (50 m ↑)		P.S. 3 (200 m ↑)	
	Day	Night	Day	Night
Numbers	1096	1482	693	1170
Volume (cm^3)	0.106	0.167	0.072	0.133
Dry weight (mg)	18.26	38.34	12.58	23.19

3.3 Seasonal variation

In the same way as there are seasonal periodicities on land, becoming more marked usually as one moves away from the equator, so there are seasonal periodicites in the seas. However, the reasons may be rather more complex. Phytoplankton responds to the increased light and

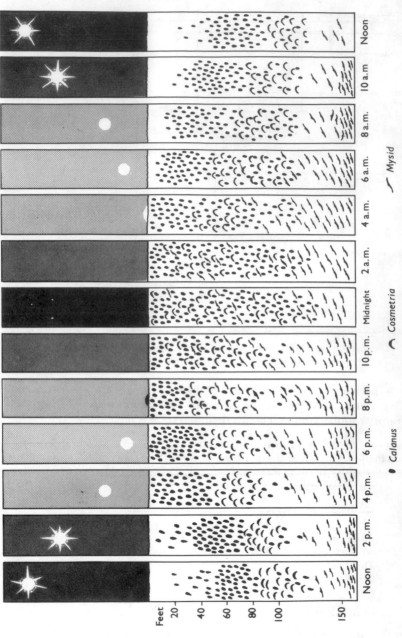

Fig. 3-2 Vertical migration of plankton (after RUSSELL and YONGE). Note that the boundaries between the three species becomes clearer with decreasing angle of the sun. The species are almost evenly distributed at midnight.

● Calanus (Cosmetria ⌐ Mysid

temperature of the spring months and multiplies rapidly as do land plants. Since there is then a much greater food supply available there is a subsequent expansion of the herbivorous zooplankton population and, after a further brief interval, an expansion of the carnivorous zooplankton population. Often the breeding time of plankton-feeding fish, or fish having plankton-feeding larvae, can be correlated with the increase in plankton.

An established thermocline (see p. 13) will form an effective barrier to nutrients passing across it. There is thus a dwindling supply of nutrients available to sustain the photosynthetic activities of the phytoplankton leading to a decrease of the phytoplankton and zooplankton populations. A change of atmospheric conditions, e.g. the onset of seasonal winds, causes the breakdown of the thermocline and nutrients formerly trapped below are now mixed with the upper layers with a resulting increase in the phytoplankton and zooplankton populations. Figure 3–3 shows the differences in seasonal fluctuations at different latitudes. This situation is of course highly simplified and, as we have seen, many other factors can influence seasonal fluctuations either directly or indirectly.

The double peak of plankton shown for temperate latitudes in Fig. 3–3, is due usually to the cycle of (1) nutrients stored in winter due to the inactivity of the phytoplankton, (2) an increase in temperature and light in spring causing the first peak, which uses up much of the overwintered nutrients, (3) a period of atmospheric stability 'locking up' nutrients by non-mixing of the water column, (4) depletion of plankton as nutrients are used up, (5) a period of atmospheric instability causing mixing of the water column and making available further nutrients, (6) a second plankton peak as the plankton uses the nutrients, and (7) the winter trough as low temperatures and low light incidence suppress the activity of the phytoplankton.

Where seasonal variations occur in the tropics, causes must generally be looked for other than the direct influence of summer and winter. Some areas are affected by monsoons, often accompanied by heavy rains. This results in a substantial run-off from the land, bringing nutrient material with it. This is sufficient to cause a marked increase in the plankton (and fish) population, seasonal in nature, but not dependent on light or temperature.

Some parts of the world, particularly around the Indian Ocean, are characterized by seasonal changes in wind directions; for one part of the year there might be persistent off-shore winds, while for another part there might be persistent on-shore winds. Persistent off-shore winds result in transport of surface water coming from below the euphotic zone, a process called upwelling (see p. 8). This constant replenishment of nutrients sustains a large phytoplankton population with concomitant large zooplankton and fish populations.

With a persistent on-shore wind the reverse will apply; the nutrient-rich water will be kept at its below euphotic zone level by the piling up of

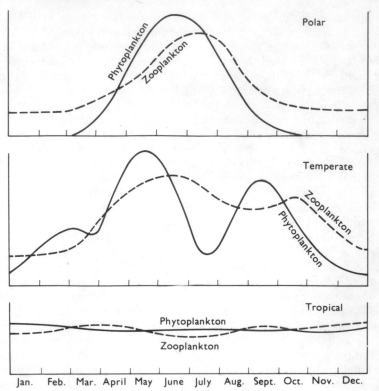

Fig. 3–3 Diagrammatic representation of the seasonal variations of phyto- and zooplankton in polar, temperate and tropical latitudes. (After FRASER, J. H.)

surface water against the coast. Without replenishment this surface water becomes nutrient poor, with resulting poor plankton and fish populations.

It should be appreciated that these are simple descriptions of what are, in fact, complex phenomena.

3.4 Spatial variation and the concept of indicator species

Apart from a few isolated inland seas, the oceans are continuous. Thus, theoretically, oceanic and atmospheric phenomena occurring in one area can eventually affect other areas. Similarly, theoretically, as the seas are continuous, any animal living in them could become evenly distributed throughout. This does not of course happen.

The upper layers of the seas have a well-established current régime, affecting the surface waters. The deeper parts of the seas also have well-established systems of water movement. As an example, the dry climate of the Mediterranean causes a higher than average rate of evaporation. Loss of water and cooling results in greater density of sea water. This relatively high density water sinks, and flows downwards and outwards over the bottom of the Straits of Gibraltar into the Atlantic westwards as far as 70°W and beyond. The base of this 'tongue' of water widens to take in the Canary Islands to the south and the coast of Spain and the lower Bay of Biscay to the north. Sometimes this water extends upwards past the west coast of Ireland and up and around the coast of Scotland, carrying with it its distinctive plankton population (see below). However, the physiology of an organism which permits it to live in one water system might not permit it to live in another. An extreme example can be seen in the transition between the warm saline water of the Gulf Stream and the cold, less saline water of the Labrador Current where, over a distance of a few metres, there can be a temperature difference of up to 11°C. This difference is an effective barrier preventing cross-migration of the plankton communities.

Where the factors separating water masses are slight, much of the zooplankton will be able to pass freely from one water mass to the other. This can be repeated, but the cumulative effect of the small barriers eventually brings about a condition where the animal reaches the limit of its viable range.

So, in extreme cases, zooplankton communities are quite sharply divided, while with less-defined boundaries separating the water masses, the change from one community to another may occur gradually over many hundreds of metres vertically, or many hundreds of miles horizontally.

Species of zooplankton, as with all animals, can live and thrive only in conditions which suit their particular physiology. An animal can be widely adaptive and live in a wide range of conditions; or it might be restricted to a strictly limited range of conditions. It follows then that a given zooplankton species with a known range of conditions can be used to predict the conditions of the water from which the sample was taken. This leads to the concept of 'Indicator Species' illustrated by the following examples.

During the summer months plankton samples taken off the north-west of Scotland can contain animals indicative of warm water. Closer examination reveals that they are members of the Lusitanian plankton in water which has originally come from the Mediterranean (see above) and which maintains its identity for a long distance. As the water cools and loses its identity as a discrete water mass, the animals die. The reverse also applies, and the limits of cold water currents running into the tropics, e.g. the Benguela Current, can be recognized by the different plankton populations.

Some zooplankton animals are to be found only in coastal or shallow waters; others indicate oceanic waters. At times normally deep-water zooplankton can be taken near the surface, indicating an upwelling movement. The reverse can apply.

It follows that although the seas are continuous, the zooplankton is not distributed evenly throughout, but conforms to certain patterns depending on the physiology and adaptability of the animal concerned. An experienced plankton worker can glean much useful information from a plankton sample from an unknown source by making an assessment of what the animals indicate.

3.5   Zooplankton communities

In general terms a community is the name given to a naturally occurring, multi-specific group of organisms living in a common environment. If neighbouring water masses differ only slightly, then the plankton communities of both can be expected to be similar. Where adjacent water masses differ appreciably, then the plankton communities will also differ appreciably.

The different zooplankton species in any given water mass will show varying degrees of hardiness and adaptability. In any one water mass, for some species the environment will represent the optimum; for others the conditions may mark the extreme of tolerable conditions. Animals will therefore occur in their greatest numbers towards the centre of that area where conditions are most suitable. As one moves farther away from the most suitable area the density of the population gradually becomes less as conditions become less suitable until, finally, they disappear altogether. Figure 3–4 illustrates this diagrammatically; note the isolated area, Y. A large consistent ocean current produces a series of eddies which become separated eventually from the main current and persist for some time as isolated bodies of water with quite different characteristics from the general water mass surrounding them. This is the case with the Gulf Stream (Fig. 3–5). Such an isolated body of water can be quite large and of course contains a plankton community similar to that within the main current and different from that of the water mass surrounding it. The area Y in Fig. 3–4 illustrates one of these isolated eddies.

Communities of zooplankton can be used to indicate certain conditions in the same way as individual species. For most of the year the Zanzibar Channel between Zanzibar and the mainland shows a typical shallow-water, warm-water, coastal zooplankton community containing many larval forms of coastal-dwelling adults. During a short period of the year a very different sample can be taken showing a community of typical continental slope forms: a clear indication of the entry of oceanic water into the Zanzibar Channel.

South of Newfoundland, around the Sable Island area, samples of two quite different zooplankton communities can be taken, one being typically warm water, originating as part of the Gulf Stream, the other being typically cold water, originating from the Labrador Current. Thus a plankton sample from this area would show very readily which body of water was being sampled.

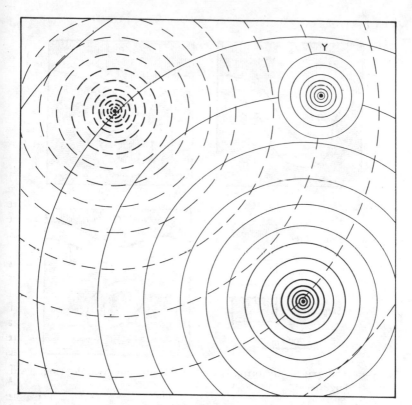

Fig. 3–4 The decrease in zooplankton as the distance from the area of optimum conditions increases. Note isolated eddy population Y.

3.6 Giant larvae

The major part of the meroplankton consists of larval forms, the great majority of which show no unusual characteristics. The benthic or nektonic adults breed at about the same time every year, the eggs and larvae passing through the appropriate number of stages, the sizes attained by the different stages being more or less the same, and the change, or metamorphosis, to the adult form occurring at more or less the same stage and size in the life history. This is not the case, however, with the so-called 'giant larva', a form not unfamiliar to the sampler of oceanic plankton.

These giant larvae are larval forms which, although usually recognizable and identifiable with normal-sized larvae, continue to grow and develop well beyond the size of the normal larvae. The important point is that these giant larvae, be they *Tornaria* larvae of hemichordates, *Auricularia* larvae of holothurians, brachiopod larvae or crustacean larvae, are almost invariably taken in oceanic conditions over great depths of water.

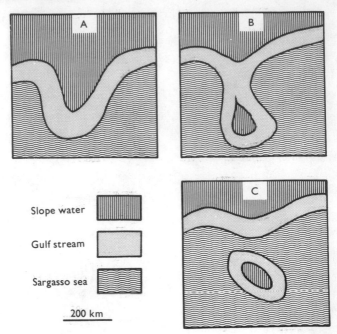

Fig. 3-5 The stages in formation of Gulf Stream cyclonic ring. (After Fuglister.)

The theory to explain them which is held generally is as follows. Adult forms live in shallow water, the planktonic larvae being found normally in coastal plankton communities. These animals show diurnal vertical migration in varying degrees, and when the settling stage is reached these larvae normally have no difficulty in finding the bottom. If, however, deep water is close to shallow water, it is possible for the larvae to be carried out to deep water. This means that when the settling stage is reached the bottom is inaccessible. This must result in the death of many larvae, but some continue their growth, albeit at a slower rate, developing into giant larvae.

These giant larvae show adaptation to a prolonged pelagic existence not shown in the normal larva. Even more interesting, some of these giant larvae develop sexual characteristics, e.g. the *appendix masculina* in some crustacean larvae and gonads with advanced stages of gametes in amphioxides larvae.

The giant larvae continue the diurnal vertical migration pattern until they either die or reach shallow water again. On reaching shallow water, contact with the bottom is established and this results in a rapid metamorphosis to the adult form, which is somewhat larger than would result from a normal metamorphosing larva.

4 Zooplankton in the Ecology of the Oceans

4.1 Its place in the food chain

It has been stressed in this book that all animal life is dependent on the photosynthetic capabilities of green plants. In the sea the green phytoplankton are the primary food producers and the next stage is the transfer of energy of this primary food through the food chain. The zooplankton fulfil a vital role in this since they are the intermediary step between the phytoplankton and larger marine life.

The vast numbers of the zooplankton make up a community in themselves and among them there are carnivores, scavengers and filter-feeders in addition to the herbivores; but as in any balanced community, the herbivores are the dominant animals in terms of their biomass.

The Copepoda are the dominant form of marine zooplankton, forming usually some 50% to 80% numerically of the zooplankton. Very few copepod species are exclusively carnivorous, the great majority being omnivores, feeding directly on the diatoms and dinoflagellates. Another very important crustacean group is the Cladocera; these are exclusively herbivorous and feed on the smaller phytoplankton. A group of much larger animals is the euphausiid crustaceans, which includes krill; these are sufficiently large to be the direct food of the largest organisms known, thus forming one of the simplest food chains, phytoplankton —euphausid—whales (Mystacoceti).

Not many food chains are as short as this. Usually larval fish, plankton-eating fish and small cephalopods feed directly on the zooplankton. Clearly it is of great importance to fish larvae, whose powers of movement are limited, that there should be a good food supply available. The breeding periods of many fish having plankton-feeding larvae are coincident with periods when large concentrations of plankton can be expected to occur. The larval fish usually selects individual organisms; but many of the small adult plankton-feeding fish have specially developed gills and gill-rakers to enable them to filter out the plankton. Herring and mackerel can feed in this way; so also does the basking shark. The situation is rather different with the plankton-feeding whales. Here a mouthful of water plus plankton (krill) is taken, and the tongue raised, thus forcing the water out through the baleen plates, which act as the filtering agent, the krill being retained and subsequently swallowed. Plankton-eating fish are the prey of various predators, e.g. king-mackerel, tuna, barracuda, which can be regarded as being at the end of the food chain.

Many benthic animals rely directly on plankton for food, e.g. barnacles and tube-worms. A particularly good example is coral. A coral reef is the basis of a complex community. The coral polyps feed on plankton and, in turn, are food for a variety of organisms. In addition it is the physical home for all manner of organisms.

Zooplankton can also be of indirect food value. Vertical migration and the Ladder of Migrations have been discussed elsewhere (p. 20) to show how food energy made available by the phytoplankton can be brought down to the depths of the sea. The animals themselves which distribute the food energy are thus available directly as food but, in addition, they can contribute indirectly to food chains. Mention has been made of how a degree of layering is present in the oceans, as for instance with thermoclines (p. 13). It is often possible to detect concentrations of zooplankton fairly coincident with the physical layers. There is some evidence to show that the excretory products of the zooplankton could provide the necessary support for a population of protozoans and small metazoans. These in turn could support some of the small larval forms which can be found at these depths; and again we follow the food chain until we reach the large predators.

In sum then it can be said that the zooplankton is the vital link through which the food energy of the phytoplankton is made available to the remaining animal life in the sea.

4.2   Zooplankton and the fisheries

Generally, fish of commercial value can be divided into two types: demersal and pelagic. Demersal fish such as cod and plaice live on or near to the sea bottom and are caught with trawls, or bottom long-lines. Pelagic fish live in mid-water and are caught with drift nets, pelagic trawls, and lines; these include herring and mackerel. Virtually all commercial fish hatch from pelagic eggs, a notable exception being the herring. Thus virtually all commercial fish start life in the plankton and can spend up to two or three months in it after hatching. Since these eggs and larvae will contribute to future exploitable fish stocks, their position in the plankton community is clearly important.

A fish egg has no positive defence mechanism even though its transparency can render it difficult to see. The size range of fish eggs is usually between about 0.5 mm and 3.0 mm; the smaller eggs are usually the most numerous, but they are liable to be eaten by a greater variety of carnivores, including some copepods and chaetognaths. A number of zooplankton carnivores, particularly ctenophores and medusae, are capable of feeding on fish eggs of all sizes should the opportunity occur. Strangely, an organism which can devour large numbers of fish eggs is

Noctiluca (Fig. 4–1). This dinoflagellate, although only up to about 1 millimetre diameter, is capable of ingesting fish eggs almost its own size though quite how is not yet fully understood.

It can be seen that if a particular year for, say, ctenophores is very good, this could result in a heavier than usual mortality in the fish eggs, with fewer fish becoming available for exploitation in subsequent years.

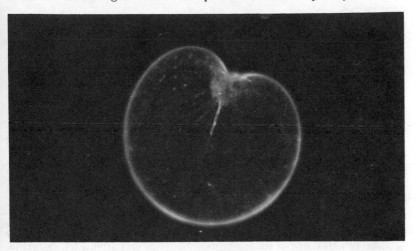

Fig. 4–1 *Noctiluca*

After the eggs hatch all the hazards of predation are still present and, when its food reserves are used up, the larva has the problem of hunting its own food. Small fish larvae do not have the energy resources to be very active in pursuit of food, or to range far and wide for it, so it has to be easily accessible.

Depending to some extent on the temperature of the water, fish larvae are planktonic for several months and, throughout this period, depend on the plankton population, of which they form a part, for their food. A marked change in the environment, apart from having a direct effect on the larval fish population, can affect the plankton in general. If the environmental change is for the better, an increased plankton population means more food available for fish larvae and a higher survival rate, and conversely.

Brief mention was made above of certain predators being common at some periods, with the attendant hazards to fish eggs and larvae. Swarms of non-predators can also have a pronounced effect. Salp populations can increase to 'swarm' dimensions over a very brief period. These animals are filter-feeders, extracting the nanoplankton and small phytoplankton from the sea. Clearly such swarms will extract large quantities of food and will thus deprive other planktonic animals. These other planktonic animals would constitute the food of fish larvae, thus such swarms would, albeit indirectly, deprive fish larvae of their food. As would be expected, since the fish eggs and larvae form part of the plankton community, they would be affected by any marked variations in other parts of the community, and these effects would be carried through to the adult populations.

Plankton can also affect fisheries directly. The swarms of salps mentioned above may be near the bottom and not visible from the surface. Should these swarms be in fisheries areas two things can happen. Firstly the fish tend to avoid such swarms and therefore fishing operations become non-productive. Secondly, huge numbers of the salps are caught in the nets which, apart from making the nets difficult and unpleasant to handle, cause physical damage to the nets by their sheer weight when lifted from the water. This would apply even more to swarms of jelly-fish.

With particular reference to herring, some years ago a theory of exclusion was put forward, in essence as follows. Herring feed on the zooplankton, particularly the copepods; therefore together with any large population of copepods there will be a good population of herring. Some phytoplankton species have the ability to form large, dense swarms. Herring would avoid these swarms, and are, in effect, excluded from the areas of water occupied by the phytoplankton swarms.

Some practical application of this knowledge was attempted. Herring fishermen were provided with a simple instrument for sampling the plankton, catching the plankton on a disc. If the disc were green or brown, this would indicate phytoplankton, and no fishing would take place. If, however, the disc were red or pink, this would indicate zooplankton, and the fishermen would put down their nets. Sometimes this worked, sometimes it did not. Such problems are never as simple as they might first appear, but it does give some indication of how a knowledge of plankton can directly affect fishing techniques.

5 How Much and How Many

5.1 Collecting

An important point which must be stressed at the outset is that no method has yet been devised that will collect the whole range of zooplankton organisms in any sample of water. This is not really surprising when one considers that the size range of plankton is from thousandths of a millimetre up to over two metres. It is better to regard plankton in the light of a spectrum: any given piece of equipment can sample one colour, with a little overlap. The collecting apparatus must be chosen which best suits the size range of organisms to be studied and more than one device used if the size range is large.

There are three main collecting methods—by bottle, by pump and by net. Bottles are caused for capturing the smallest organisms and have usually a capacity of between 5 and 20 litres. They are made generally of an inert plastic. In the laboratory the bottle water is filtered to remove the organisms. The main advantages of bottles are that they can be lowered to any depth with reasonable accuracy and the organisms captured come from a known volume of water. Disadvantages are that bottles are only suitable for the smallest organisms.

A pump is normally situated on the boat deck, but sampling can be equally well carried out from a pier or similar structure. An inlet pipe is lowered into the sea and an outlet pipe directed into a suitable receptacle. The pump is usually centrifugal. Small organisms can pass through it intact, but larger organisms may be mutilated. The receiving tank is of a known volume so that, after filling, the organisms can be filtered off and related to a known volume of water. When sampling larger volumes a net of suitable mesh is put into the tank with only the rim of the net above water and the outlet pipe directed into it, suitable arrangements being made for the overflow. This prevents any damage to the organisms such as would occur if the net were suspended in the air and the water passed through. A meter on the pump records the volume of water pumped. Advantages are the ease of use, accuracy of sampling depth and easy measurement of the volume of water sampled. Disadvantages are the potential mutilation of the organisms, the limited depth from which it can operate and the relatively small diameter of the inlet pipe. This last precludes the capture of any larger, active organisms since they can evade easily the slight suction at the mouth.

A net in some shape or form is by far the commonest method of

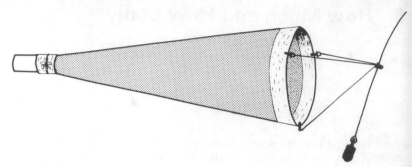

Fig. 5–1   Simple plankton net. The weight or depressor takes the net below the surface and a rigid net ring holds the mouth open. Plankton is caught in the bucket at the end when the net is raised.

plankton sampling. The netting must be of accurate and fixed dimensions and of durable material. Netting today is mostly nylon or terylene, though in some of the so called high-speed samplers the nets are made from stainless steel or Monel metal. Mesh apertures can be as small as 15 μm, or, for catching larger zooplankton, 2 centimetres or more.

In its simplest form a plankton or tow net is as shown in Fig. 5–1. In some nets the mouth is non-rigid but present inclined planes to the water so that forward movement opens the mouth automatically.

Plankton that is too large to pass out through the mesh apertures is retained and is concentrated in the bucket. Since, in practice, a fair proportion of the sample remains sticking to the sides of the net it is very important to wash down the net before removing the bucket. The bucket, plus plankton, is then detached and the contents can be fixed—or treated in whatever way is needed.

There are many variations on this basic scheme; but various considerations will apply to all cases. When towing a plankton net the object is for water to enter the mouth and to be filtered through the mesh at the same speed as that at which the net is towed through the water, thus getting 100% filtration. At high speeds the filtering surface offers too much resistance to the passage of the water and a static cone of water develops in the net, water being diverted outside the net ring. Consequently, there is no effective filtration at all. Intermediate speeds will achieve differing degrees of filtration.

The area of the mouth aperture has a fixed relationship with the area of filtering surface represented by the total mesh aperture area of the conical nylon net. It should be appreciated that if the plankton is particularly abundant, or if there is an abundance of gelatinous organisms, the meshes will tend to become clogged, reducing the actual filtering area and thus reducing the efficiency of the net.

It follows that permutations of mouth aperture, net length, mesh aperture and towing speed all affect the efficiency of the net and the size of the organisms it retains. In practice the portion of the zooplankton spectrum to be sampled determines the design of the net and the conditions under which it is used.

It should now be clear why a plankton sample cannot give a valid estimate of the *total* zooplankton, but only a reasonable estimate of that part of the zooplankton spectrum which the apparatus is capable of sampling. This is why any estimates stating the total zooplankton in the sea which have been based on samples from a single piece of collecting apparatus are not valid. Table 5–1 illustrates some of the errors which can result. The siphonophores are largish animals and will be caught by all the nets so the numbers do not vary greatly.

Table 5–1   The variability of the no. per m³ of some different plankton animals according to the mesh aperture size.

	Mesh aperture size				
	$1000\,\mu m$	$650\,\mu m$	$300\,\mu m$	$200\,\mu m$	$70\,\mu m$
Copepod nauplii	1(?)	4	8	90	10 000
Copepods	7	20	450	780	2500
Chaetognaths	4	20	35	90	700
Siphonophores	15	20	20	20	10

Another important consideration when interpreting zooplankton samples concerns elongated animals such as fish larvae and chaetognaths. Although they may be several millimetres long, the cross-section diameter is but a fraction of a millimetre. Thus if a net with, say, 500 μm mesh apertures is used, the animals will be retained if they come up against the mesh broadside on, but will be able to pass through the net head first. Therefore only a proportion of the total animals will be caught.

In order to arrive at any quantitative zooplankton estimations it is necessary to know the volume of water which has been filtered to produce the sample. This is relatively simple with bottle or pump sampling. With net sampling a flow meter is used to indicate the passage of a known volume of water, expressed usually in cubic metres. However, in this sort of work it is rare to find situations to be as straightforward and simple as they might appear. The recorded volume will depend upon where the flowmeter is placed. The use of a flowmeter does not necessarily mean therefore that the volume of water filtered is measured accurately, but if the same apparatus is repeatedly used in the same way, then the results are directly comparable. This is an important practical point. The fact that many varied items of sampling gear are used creates a problem when the results of samples taken with them have to be related.

If one wishes to sample the plankton between, say, 1000 and 500 metres some closing device must be fitted. One of the simplest forms is shown in Fig. 5–2. The net is lowered to the required depth (nothing is caught while

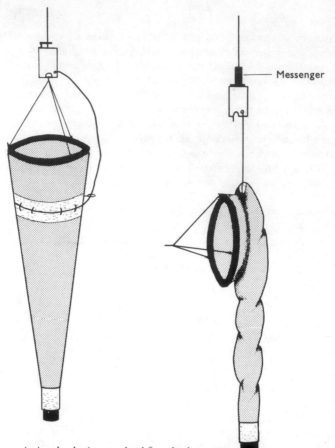

Fig. 5–2   A simple closing method for plankton nets.

the net is lowered backwards) and then hauling in begins. At the appropriate time a messenger is sent down the wire; this releases the catch, thus allowing the net to be throttled so that nothing more is caught as the net is hauled to the surface, and what has been caught is retained. In practice, this method is not as efficient as it might seem. For more sophisticated nets a closing mechanism, or indeed an opening and closing mechanism, can be worked by sonic impulses from the ship. Some nets have a built-in electric power-pack which opens and/or closes the nets. Some nets can be used to take a series of discrete samples from successive depths using such triggering devices.

There have been many refinements derived from the essentials of net sampling for zooplankton described above.

It is important to know the actual depth of a net, the Isaacs-Kidd trawl, for instance, when it is being towed horizontally at any depth. When possible, a pressure-sensitive transponder is fitted to the head of the net.

This transmits an appropriate signal which is picked up by the ship, thus indicating the actual depth at which the net is being towed. An alternative is to incorporate in the net a needle connected to pressure-sensitive bellows, the needle recording the depth on a suitably prepared surface. The main disadvantage of this is that the depth is not known until after the sample has been taken.

Catching the larger, faster zooplankton is a problem that has not yet been solved satisfactorily. Faster towing speeds of 6–8 knots mean a smaller mouth aperture, provided usually by a truncated cone at the front (Fig. 5–3). A serious problem resulting directly from towing at higher

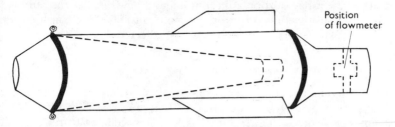

Fig. 5–3   'High-speed' tow net.

speeds is that in many nets the animals caught are badly damaged. It was, partly, to surmount this problem that the Jet-Net was devised (Fig. 5–4). In this design the net is folded upon itself for compactness, and is arranged so that the rate of exit of the water is about the same as the rate of entry. However, it will be noticed that the diameter of the intake pipe is gradually increased to where the water begins to pass through the collecting net. Since the same amount of water is passed as that which enters the net, it follows that the rate of flow is decreased. Thus at a towing speed of about 10 knots the flow of water through the mesh where the animals are actually being caught is reduced to about 1.6 knots, or about 0.8 m/s, and the zooplankton caught is in much better condition.

The number of variations on the theme of a simple plankton net is

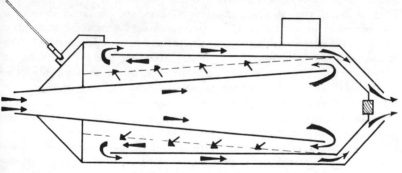

Fig. 5–4   The Jet-Net.

legion. Books have been written which are devoted exclusively to them. There is, however, one more development which should be mentioned, the Hardy Continuous Plankton Recorder. This instrument can be towed at high speeds—it has only a 1.25 centimetre square aperture at the front—and is a means of sampling the plankton along a line hundreds of miles long. A full account of this can be found in Sir Alister Hardy's book *The Open Sea, its Natural History: the World of Plankton.*

It must be stressed again that all plankton sampling devices are limited in their scope and, in any interpretations of plankton samples, these limitations must be given due consideration, and the choice of gear must relate to the range of zooplankton wanted. Likewise, after the sample is taken, conclusions drawn from the sample must take into consideration the many extrinsic factors which may have affected the sample.

5.2   Fixing and preserving

While it does not take long to obtain a plankton sample, it takes longer to analyse it properly. In addition it is usually very difficult to study samples aboard ship. The result is that samples accumulate more rapidly than they can be analysed. Some means must be found therefore of fixing the sample in as near to the living condition as possible, and preserving it so that the sample can be stored for long periods and examined at leisure. However, it cannot be stressed too strongly that living samples should be examined whenever possible. The zooplankton is composed of living organisms and, as such, display adaptations to their environment which are obvious in the living organism and would otherwise be missed. This apart, only living zooplankton reveals the beauty of line and colour of these organisms.

The fixing and preserving agent most commonly used is formalin; this is a saturated solution of the gas formaldehyde in water, which gives the 40% solution. If this is diluted with water in the ratio of 1 : 9 the resulting 4% solution is adequate for fixing. To prevent possible undesirable osmotic effects dilution should be made with sea water. The strength of the formalin is not critical provided it is not below 4%.

After fixing, that is after, say, several days, the plankton should be transferred to air-tight glass containers in about 3% formalin in distilled water. For storage and handling, several points should be noted. The container should be filled completely, avoiding the inclusion of air bubbles to limit mechanical damage during handling. Storage should be in the dark; this is some help in preserving colours for a longer period than would normally be the case, although formalin is not a good preservative for colour. Also, it helps prevent formation of undesirable compounds such as formic acid. Storage temperature should be above about 5°C, to prevent polymerization, and below about 20°C. If the sample is to be stored for any length of time—and there are many plankton collections more than fifty years old—desiccation is always a potential problem and 5–10% ethylene glycol can be added to alleviate this and the cap sealed with wax.

It should now be clear that problems concerning plankton are rarely as simple as they appear. Many zooplankton groups have calcareous shells or skeletons, e.g. foraminiferans, pteropods and echinoderm larvae. Formalin, as usually supplied, is not very pure and is positively acid. Thus the shells and skeletons, being so delicate, can be dissolved very quickly, sometimes in minutes. This is one reason why the purest quality formalin available should be used. Even so, acid breakdown compounds can be present and all formalin should therefore be buffered, with calcium carbonate or sodium bicarbonate. Borates are often used, but these tend to cause degeneration of the specimens after some time. If the calcareous zooplankton is particularly wanted, many researchers advocate the separation of this fraction by gravimetric methods and separate preservation by freeze-drying or in 70% ethanol or 40% isopropanol. If separation is not practical, the whole sample can be preserved in ethanol or isopropanol, but these do cause significant distortion and shrinkage.

The above-mentioned fixatives and preservatives are not particularly good if subsequent histological investigations are planned. It is possible to give some degree of post-fixative treatment, but this is not as satisfactory as an appropriate initial fixing. Bulk specialized fixing and preserving is not usually very practical, although Bouin's Solution—a good general-purpose histological fixative and preservative—can be used. The best method is to pick out the particular animals required for detailed examination and treat them in an appropriate way.

A mixture of 80% methanol, 10% glacial acetic acid and 10% of 40% formalin (MAF) is also a good general-purpose fixative. For general electron microscopy, suitably buffered glutaraldehyde is often used.

A golden rule for all plankton fixation is that it should be done immediately the sample is brought aboard. Zooplankton can deteriorate rapidly, especially in a warm atmosphere. In addition all samples should have an adequate informative and durable label put in with the sample at the earliest possible moment. It is advisable also to write key information on the outside of the jar, preferably on the lid. This is a great help when sorting samples, since labels in the containers can be obscured easily by the plankton.

The whole subject of fixing and preserving is specialized, and further information should be sought from specialist books; but it should always be remembered that most, if not all, zooplankton fixing and preserving agents are poisonous, with noxious vapours, and should therefore be treated with respect.

5.3   Analysis of samples

Consideration must be given to the information required from the plankton sample before the method, or methods, of analysis are decided.

Three parameters are generally analysed: volume (settled or displaced), weight (wet or dry) and numbers.

Volume measurement is easy to do, but the information gained is not of great value. Settled volume can be measured in a measuring cylinder, allowing at least 24 hours for the sample to settle. Displacement volume is also measured easily by making the sample plus liquid up to a known volume, filtering off the sample and measuring the volume of the liquid which has passed through the filter. The plankton volume is found by subtraction from the original volume. Whether the sample is gravity-filtered or vacuum-filtered can make an appreciable difference to the result.

Wet-weight, or biomass, measurement is simple but the result has limited value. The wet weight of a given sample will vary greatly depending on the method used to remove the water and the type of sample. The water can be removed by gravity, vacuum or pressure filtration; and with a large sample of gelatinous animals the water will continue to drip through for quite a long time. If a high vacuum or high pressure is used the animals can be damaged and distorted. Animals such as salps and siphonores have relatively large cavities which can retain water; this is another potential source of error. In general, wet-weight estimations are simple and can be carried out aboard ship. If consistent methods are used some useful information can result, but caution must be exercised in any interpretation of results.

The measurement of dry weight would also appear to be simple; but much confusion can be caused by the various interpretations of 'dry'. After filtration and a quick wash with distilled water (to prevent salt crystals introducing an unnecessary error), drying to a constant weight in a desiccator is regarded by many as being sufficient. However, the majority of workers are agreed that the sample should be dried in an oven, but results will vary greatly with different temperatures. Above about 70°C some combined water will be lost, volatilization of fats or a breakdown and decomposition of compounds can occur. Perhaps the temperature most suitable for drying zooplankton is 50°C.

Results from dry weights are more informative than wet weights as they are a more accurate measure of the actual amount of animal tissue. Methods are simple and they can, if necessary, be used aboard ship. It is essential to use fresh plankton; if any delay is expected between sampling, drying and weighing, the sample must be stored at about freezing point.

The above measurements, although they give some idea of the amount of total plankton present, give no information on the composition of the sample and the relative proportions of the different animals. The only way this can be done is by counting the animals in the sample. Samples can contain tens, thousands or hundreds of thousands of animals. Counting the total sample is possible in some cases but usually is quite impractical, so the sample has to be sub-sampled. Often it is advisable to take one sub-sample for the majority of the animal groups and another,

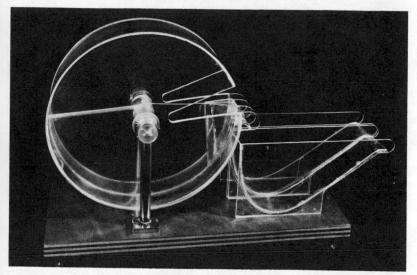

Fig. 5–5 The Folsom plankton splitter (photograph by David Nicholson).

smaller sub-sample, for the more common groups. A simple but effective way is to dilute the sample to a known volume, agitate it so that the animals are distributed evenly throughout and then remove a sub-sample of a known volume. Alternatively there are mechanical devices available, the most common perhaps being that illustrated in Fig. 5–5. Here, one half of the drum is divided into two equal lateral halves by a median septum. The drum is arranged so that the undivided half is at the bottom and the sample is poured into this. The drum is then rotated so that the divided half is at the bottom. The sample of course stays at the bottom and is divided into two equal halves by the septum. The halves are then poured out into the two containers. This process can be repeated until a suitable fraction is obtained.

Generally a dry-weight plus a numerical analysis will yield the best overall information. Apart from these general methods of analysis, something more specialized might be required, e.g. an analysis of protein, fats, elements or pollutants present.

All the finer analyses must be carried out on fresh samples in an appropriate way. They are most valuable if they can be related to the plankton community as a whole; and the only way to discover the structure of the plankton community is by unspectacular and often tedious numerical analyses.

5.4 The possibilities of laser holography

It cannot be stressed too strongly and too many times how limited in scope any particular item of zooplankton-sampling gear is when related to the whole range of zooplankton organisms. A comparatively new invention, laser holography (see LEITH and UPATNIEKS, 1965), has been used

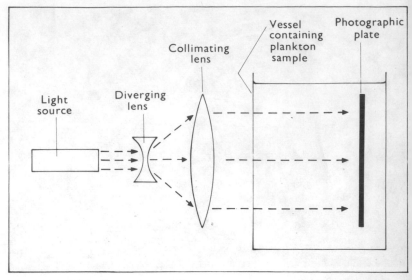

Fig. 5–6 Schematic representation of living plankton holography by laser beam.

recently in plankton work, and it does present some possibility of getting a more satisfactory picture of a zooplankton community *in situ*.

Some experiments have been conducted on the lines of the diagram in Fig. 5–6. Zooplankton was kept in the container and the photographic plate, placed about 15 centimetres from the front, was actually immersed during the exposure. A ruby laser was used, the exposure being a pulsed flash of about 60 ns. The plate was developed in the usual way and viewed through an ordinary microscope. With no image-forming lens there is no depth of focus problem. The total volume recorded was about 1000 cm^3 and any point in this volume may be brought into focus. The brief exposure was more than sufficient to 'freeze' the animals so that no problem of blurring through movement occurred, and the organisms were recognizable.

So here we have a 'frozen', three-dimensional picture of a volume of water in which there is a small population of living zooplankton, thus preserving spatial relationships and, within the limits of the photographic emulsion, recording all the different sizes of organisms present. Developments are now being considered whereby a self-contained piece of apparatus can be lowered into the sea to take an *in situ* hologram of a volume of water in which all the organisms can be observed in their proper relationship, regardless of relative size.

Of course a tremendous amount of work has yet to be done before it becomes a practical proposition; but it does hold out some hope that here there is, potentially, a piece of apparatus that will be able to sample almost the whole spectrum of zooplankton, albeit as a three-dimensional hologram and not as a physical sample.

6 Adaptation to the Environment

6.1 Structural adaptations to flotation

The environment and plankton community together form a comprehensive ecosystem and, as in any ecosystem, there are interactions between the organisms and also between the organisms and the environment. These interactions give rise to many adaptations by the organisms to a planktonic existence. The zooplankton are animals living, in effect, in a state of suspension in the sea, and show, therefore, some adaptations that help the animal maintain its state of suspension.

With no other factors to complicate the situation, the rate at which a body falls through water is related to the surface area offering resistance to the passage through the water. A sphere has the least surface area relative to its volume, so it follows that, within the size range of the zooplankton, a sphere would be the worst shape for a plankton animal to develop. This accounts for the fact that zooplankton are not spherical unless some positive buoyancy mechanism is present, e.g. oil globules and dilute fluids in fish eggs. The effect of the relationship between surface and volume is easily demonstrable with a vessel of water and a small piece of plasticine. A ball of plasticine of about 2 millimetres diameter will sink rapidly in a straight line to the bottom. The same piece of plasticine when moulded into different shapes will fall at different rates, and any differences are therefore due only to the changed shape. Many zooplankton are found with body extensions which increase their surface area, resulting sometimes in bizarre shapes.

The most common multicellular zooplankton are the crustaceans. All have an exoskeleton and are therefore relatively heavy animals. This weight is often counterbalanced by extension of the exoskeleton into many spines. A simple example is the larva of the common shore crab, *Carcinus maenas*. Other crab larvae have a pair of lateral spines on the carapace and in some species two pairs may be developed (see Figs. 1–1 and 1–2). Additionally the spines may terminate in a flat 'spear-head'. A large elongation of the anterior spine is characteristic of porcellanid crabs. An extreme development of spines is shown by *Sergestes* larvae.

Echinoderm larvae may show a large surface area/volume ratio owing to the presence of many long spines, or arms. Some have a calcareous skeletal axis, while others maintain flexibility of the arms for use in swimming (Fig. 6–1).

Radiolaria with their numerous spines have been known for many

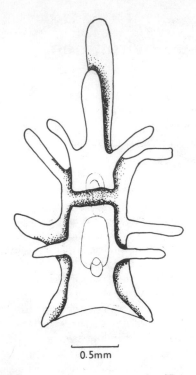

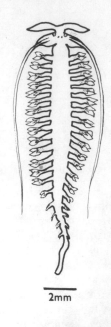

2mm

Fig. 6–2   *Tomopteris.*

0.5mm

Fig. 6–1   *Bipinnaria* larva of starfish.

years for their beauty of form. *Globigerina* presents a smooth appearance when examined as part of a bottom deposit—the Globigerina ooze—but when living in the plankton it has innumerable fine calcareous spicules radiating from the test.

Clearly, the use of skeletal extensions to improve flotation is not open to animals without skeletal structures. However, in the case of some pelagic polychaete worms a dual-purpose structure has evolved. *Tomopteris* (Fig. 6–2) has well-developed parapodia which, apart from increasing the surface area, are also an adaptation for swimming. Many polynoid larvae have an extensive covering of long setae.

Molluscs would appear to be at some disadvantage as plankton animals since most have a shell. In some cases, e.g. *Cavolinia*, the shell can be relatively massive. A common molluscan larval form is the veliger (Fig. 6.3) in which a large lobed velum has been developed. This greatly increases the body area, acting almost as a parachute, and enlarging the circlets of cilia which are used for swimming and feeding.

The copepods have developed plumose setae rather than spines (Fig. 6–4). In addition some are very colourful, and living copepods can be objects of great beauty. Unfortunately it is virtually impossible to capture such a copepod with all the setae intact.

A thin, leaf-like shape will present a greatly increased surface

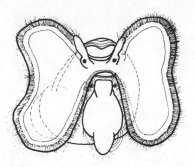

Fig. 6-3 *Veliger* larva.

|_____|
0.2 mm

area/volume ratio. We find that many copepods have developed this form, the appendages being used as stabilizers. The genera *Sapphirina* and *Copilia* are particularly well developed in this manner. Perhaps the ultimate in the development of this body form is seen in the *Phyllosoma* larva of the Scyllaridea; the name itself means leaf-body (Fig. 6-5). The

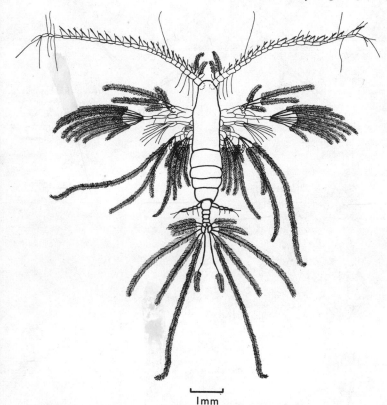

|___|
1mm

Fig. 6-4 A calanoid copepod, *Augaptilus*. (After Giesbrecht.)

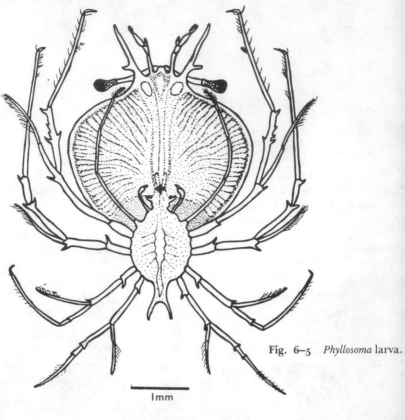

Fig. 6–5 *Phyllosoma* larva.

1mm

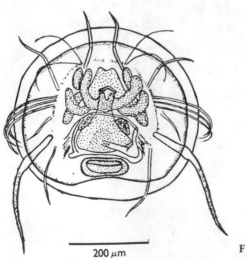

200 μm

Fig. 6–6 *Pelagodiscus* larva.

shape of medusae is well known; clearly the parachute principle is working here.

There are two forms of brachiopod larva which are not uncommon in tropical plankton; Fig. 6–6 shows one. Both have a thin, chitinous, bivalved shell enclosing the body, and the body has appendages capable of projecting from the shell. Probably they float in a horizontal position, the appendages acting as stabilizers.

Tropical waters are, generally speaking, less dense and less viscous than temperate and polar waters. One result of this is that, again generally speaking, tropical zooplankton animals are smaller, thus having a larger surface area/volume ratio, and may have more body projections than their colder water equivalents.

6.2   Other aids to flotation

The flotation aids discussed above do not contribute a positive buoyancy factor; rather they act to decrease the rate of sinking. Some planktonic animals have developed positive buoyancy aids, and this applies particularly to soft-bodied animals.

The most common positive buoyancy device is the actinopterygian fish swim-bladder. This is a gas-filled sac, homologous with the lungs, containing usually a mixture of oxygen, carbon dioxide and nitrogen. The net effect of it is to counterbalance the weight of the heavier tissues of the body so that a state of neutral buoyancy is achieved. Many bottom-living fish have lost the swim-bladder by the time the metamorphosed juvenile has sunk to the bottom, but it is significant that the planktonic larva does usually have a swim-bladder present.

Gas-filled floats have been evolved in various forms in the coelenterates. *Physalia*, the Portuguese Man-o'-war, is one of the better known of the larger zooplankton animals. The arrangement of its colony is such that a large oval and coloured contractile float is developed at the top. Analysis of the float gas shows the presence of oxygen, nitrogen, argon and carbon dioxide, comprising about 14.5, 74.5, 1.1 and 0.4% respectively. Most surprising is the presence of carbon monoxide, usually about 9% of the total, but sometimes more than 20%.

Two other common surface-living forms which have gas chambers are *Velella* and *Porpita*, the former having the dorsal surface produced into a sail-like structure.

Most siphonophores live below the surface, so in those groups having a pneumatophore some means of regulating the gas in it must be present. *Nanomia bijuga* is reputed to be able to ascend for about 300 metres in an hour; this means that in order to have an inflated pneumatophore at the bottom of this vertical range, the gas content must be maintained against a diffusion gradient of 30 atmospheres. Regulation here is by an apical pore controlled by a sphincter muscle. Analysis of the float gas in this species shows between 80% and 90% carbon monoxide.

An interesting point to note is the connection between physonectid siphonophores and the deep-scattering layer. These siphonophores are very common and the presence of so many gas-filled pneumatophores to form density interfaces is well suited to being detected by depth recorders. Bathyscaphe investigations have revealed the connection between the deep-scattering layer as revealed by precision depth recorders and the presence of innumerable physonectid siphonophores at exactly the depth indicated. They are proving to be so common that one investigator speaks of '. . . a zone of mid-water predators—a living net—stretched across the world's oceans'.

Ianthina, a pelagic gastropod which feeds on *Velella*, has an unusual floating mechanism. No internal gas chamber is present, but it 'blows' a series of bubbles which are viscous in nature so that they stick together to form a raft of air bubbles to which the *Ianthina* is attached.

A useful substitute for contained gas is a concentration of a tissue, or food reserve, which is only moderately less dense than sea water. The majority of marine fish lay eggs which float. Since they are passive they must have a positive buoyancy. This is achieved by including within the egg membrane a yolky food reserve, scattered oil droplets or possibly one or two large oil globules and dilute body fluids. The size and disposition of the oil globules can be important in the identification of the eggs. At a later stage the young fish itself may concentrate fatty or oily tissue which helps buoyancy.

Many crustaceans contain a significant quantity of oil or fatty tissue in the body; this is particularly true of the herbivores and omnivores. When euphausiids are preserved many oil globules can often be seen, having leached out from the bodies. Deep-sea copepods always seem to have a larger than usual amount of oil in the body.

A high proportion of the marine zooplankton is gelatinous in nature, e.g. medusae, ctenophores, some molluscs, thaliaceans, some nermerteans. It has been suggested that the gelatinous tissue had a buoyancy function, and subsequent investigation showed that this is indeed the case. The mechanism involved is an alteration of the ionic balance of the body fluids by the partial exclusion of heavy sulphate ions, these being replaced isosmotically with chloride ions. Fig. 6–7 shows the sort of result which can be expected.

One of the delights of the study of plankton is the observation of live animals, and few are more fascinating than living cranchid squid. The majority of squid could never be considered as helpless planktonic animals, but the cranchids reasonably can. It is not uncommon to capture individuals several inches long in deep water plankton samples (they are essentially inhabitants of the deeper waters) and if in good condition can be observed alive. Many hang head downwards in the water in a condition of neutral buoyancy. It is only recently that the buoyancy mechanism has been understood. It has been shown that the coelomic fluid has a specific gravity of only about 1.01 (sea water being about 1.026) but is nevertheless

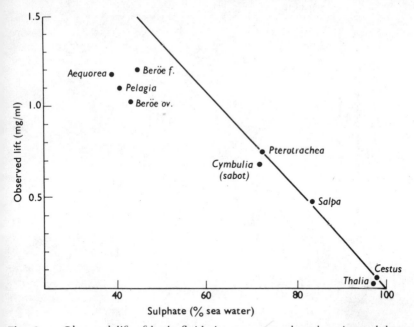

Fig. 6–7 Observed lift of body fluids in sea water plotted against sulphate concentration. The line shows the lift given for sea waters in which sulphate has been replaced isosmotically with chloride. The abcissa represents sulphate concentrations of body fluids as a percentage of sulphate concentration of sea water. (After Denton and Shaw.)

isosmotic with sea water. This gave the clue that the ammonium ion was involved, and subsequent analyses showed the coelomic fluid to consist largely of ammonium chloride solution, ammonium ions being present in the high concentration of about 9 grammes per litre. The source of the ammonium is easily explained since, in the metabolism of the squid, the nitrogen excreted from the breakdown of proteins is in the form of ammonia.

The use of isosmotic solutions of ammonium chloride as a buoyancy mechanism has been found in other cephalopod groups, in a crustacean larva and also in the dinoflagellate *Noctiluca*.

Apart from those animals having neutral or positive buoyancy, the problem of the prevention of sinking is an ever-present hazard to the zooplankton, which must be overcome. In some animals which are almost neutrally buoyant little effort is required, while others need to swim vigorously. There is a striking difference between *Sagitta enflata* and *Creseis acicula* when kept alive together. *S. enflata* (Fig. 6–8), a common warm water chaetognath, is transparent, soft bodied and almost neutrally buoyant so that an occasional burst of swimming activity is all that is necessary. *C. acicula* (Fig. 6–9), a thecosomatous pteropod, has a relatively

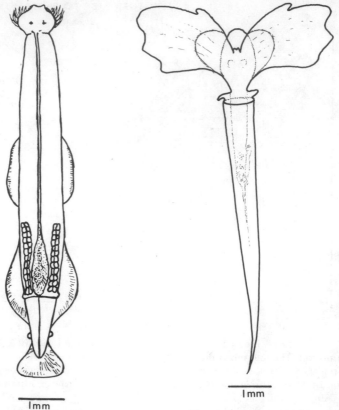

Fig. 6–9 *Creseis acicula.*

Fig. 6–8 *Sagitta enflata.*

heavy and cumbersome shell and needs to swim very vigorously by means of its 'wings' in order to remain in mid-water. When kept in a tank *C. acicula* sinks through the water rapidly when swimming ceases.

When kept in an aquarium with static water the majority of the zooplankton will eventually sink to the bottom and die because of the lack of movement of the water.

6.3 Colour and transparency

Zooplankton is the intermediate stage of many food chains, and as such is heavily preyed upon by the larger zooplankton organisms or by animals of the nekton. Protective coloration, or, in many cases, lack of coloration, has been evolved to afford some protection from predators.

Being invisible is an efficient means of protection, and varying degrees of transparency are found in most of the zooplankton. The crustaceans are not as able to adopt transparency in view of their exoskeleton, but even so some do achieve a remarkable degree of transparency. Of the smaller crustaceans the male *Copilia*, a tropical pelagic copepod, is so transparent that the whole anatomy is beautifully displayed, with the nervous and muscle systems laid out as if in a diagram.

The larger crustaceans of the plankton, particularly the larval forms such as are common around the British coast, have some degree of transparency, but usually have a pattern of pigment spots over the body; these are an important aid to identification. In the open ocean, particularly in the tropics, it is easy to attract numerous animals by hanging a light near the surface of the water on a dark night. Among these animals will be stomatopod larvae several centimetres long and, apart from a beautiful green-blue pigment behind the eyes, perfectly transparent.

Varying degrees of transparency are shown by the zooplankton which have soft gelatinous bodies. Some cranchid squid are almost completely transparent. I have seen what were, to all intents and purposes, a pair of disembodied eyes floating in a plankton sample which, on inspection, turned out to be such a squid. Perhaps the most transparent animal is the pseudoconch of *Cymbulia*. Even after immersion in formalin it is still virtually invisible when put in sea water; before preservation it is quite invisible. Preservation will render all the animals more opaque than they are in life, so a proper appreciation of the transparency of zooplankton can be realized only by examining living material.

Some animals, although generally transparent, are tinted different colours. Thus some salps appear pale blue; *Velamen* and *Beröe*, two ctenophores, are often pale pink. Some animals, particularly crustaceans, are generally transparent, but due to a finely sculpted cuticle or exoskeleton, have a prismatic effect on the light. Outstanding in this respect is *Sapphirina metallina*, a leaf-like pelagic copepod. If living specimens are put into a white bowl they will swim around quite vigorously, rocking or rotating on the long axis. Although transparent, their exoskeleton is such that white light is split into its spectral colours so that their progress through the water is marked by periodic flashes just like a diamond flashing.

A much more positive colour is seen both in zooplankton living very close to the surface and in the deeper waters. The Neuston Net, developed only recently, has enabled the plankton in the top few centimetres of the sea to be investigated. In the many different animal groups represented in the neuston, one feature common to many of the animals is a bright blue colour. This is produced by a carotenoid pigment, astaxanthin, linked to a protein, with a maximum absorption at about 650 nm when extracted with water and at about 475 nm when extracted with an organic solvent. The latter destroys the protein linkage. Nobody has yet advanced a completely satisfactory explanation for this colouring, but it is generally accepted that it constitutes some protection from potentially harmful radiation, including the ultraviolet, which penetrates the upper layers of the water.

With increasing depth, transparency and some silvery coloration are common until about 500 metres is reached, below which it would seem that a deep red cryptic coloration (see p. 17) is the most common in the plankton. This is present in many animal groups, e.g. crustaceans, medusae, chaetognaths and nemerteans, but appears to be uncommon in fish, although these last are generally not to be considered as plankton.

6.4   Bioluminescence

This phenomenon is widespread throughout the living world, generally in animals, but also to be found in the lower plants, particularly the marine groups. Most, if not all, the animal groups present in the marine environment have representatives which are capable of luminescence, either spontaneously or when stimulated. Strangely, perhaps, one group of the Crustacea which has been given the generic name of *Lucifer* has not been demonstrated to be luminescent.

Generally speaking there are three ways in which luminescence is produced; (1) continuous production of light by symbiotic bacteria, (2) discharge of a luminous secretion as a luminous cloud into the surrounding water, and (3) intracellular luminescence originating from special organs, the photophores. In many larger organisms the light-generating glandular cells of the photophores are backed by a reflector, screened by pigment and have a lens to concentrate the light produced. Where light is produced by symbiotic bacteria, the bacteria are then sited in specific areas and some, or all, of the areas can be 'extinguished' by the host animal (usually a fish) by a flap not dissimilar in action to an eyelid.

Often one reads of tropical seas being a 'sea of fire' at night, but this appearance is generally the exception rather than the rule. In any case, it is not due to diffuse luminescence throughout the water but by discrete sources of luminescence originating from very small phytoplankton and zooplankton which are so numerous that an impression of diffuse continuity is given. Usually, in these conditions, the luminescence results from a stimulus; the bow-wave and wake of a boat can be highly luminescent. I have seen a 'sea of fire' situation in the Malacca Strait when dolphins were easily discernible by their silhouettes in the 'fire'. Beyond coastal and shallow waters, luminescence is seen usually as discrete light sources, some small but others, such as that of a *Pyrosoma* colony several metres long, can be very large.

The light of most pelagic animals is blue, predominantly within the 460–490 nm section of the spectrum. Some animals, such as some polynoid worms and ctenophores, produce light between 510–520 nm, and appear blue-green to green. Although these are by far the commonest wavelengths noted, there have been reports of photophores emitting yellow, pink and red light. However, some of these may have another explanation. Very often, particularly in squid, the chromatophores are highly reflective, and such chromatophores can appear to be emitting whereas, since they are 'passive' organs, they are only reflecting light appropriate to the chromatophore pigment.

There are several different mechanisms involved in the production of light. In the first the basic mechanism is enzymatic oxidation of a substrate. In the second there is a substrate precursor which is first activated and then oxidized. In the third, molecular oxygen is not required, the reaction being based on peroxidation. Fourthly, immediately preceding the oxidation, a reductive process is involved. In contrast to the other mechanisms, this provides continuous light emission and is found in bacteria and fungi. A fifth mechanism has been suggested whereby all the necessary reactions occur, but instead of the flash occurring it is put into store, so to speak, to be triggered off when required.

It is interesting to note a luminescent phenomenon which would fall into this fifth category. From individuals of the hydromedusan genus *Aequorea* there has been extracted a so-called photoprotein called aequorin. Luminescence is emitted from aequorin only in the presence of calcium ions. Aequorin has been used as a tool by neurophysiologists to detect the presence of ionic calcium during various states of nerve excitation; the calcium can be detected and measured in 0.1 μm concentrations.

How bright is the light generated by general luminescence? Certainly the flashes are readily discernible; in some parts of the world fireflies are used in lanterns, heaped together in glass containers. Additionally one must of course differentiate between what is discernible by man and what is discernible by organisms living in the sea.

The intensity of sunlight is variable with atmospheric conditions and latitude, but a reasonable figure for light intensity at the surface of the sea is 10^5 μW/cm^2 (0.1 watts per square centimetre) (see Fig. 2–10). A fully dark-adapted human retina can perceive light intensities of about 10^{-8} μW/cm^2. Some of the brightest luminescent flashes are in the region of $10^{-3}$$\mu$W/cm^2, as detected at a distance of a metre, and are thus easily perceived by the human eye. The light intensity of flashes emitted by single planktonic protozoans has been measured, and is in the order of 10^{-9} μW/cm^2 at a distance of a metre. Due to their rather different structure, or quality of structure, the eyes of deep-sea animals, particularly fish and squid, are believed to be 10 to 100 times more sensitive to such flashes than the human eye (Fig. 2–10). Their visual pigment is also most sensitive to blue or blue-green light; the human eye is most sensitive to yellow-green (550 nm) during the day and green-blue (505 nm) when dark-adapted.

Luminescence is such a widespread phenomenon that one must assume it to have an important function; but, with one or two exceptions, these functions have not been resolved satisfactorily. It is reasonable to suppose that extracellular luminescence is, in many cases, a defence mechanism. Acanthephyrid shrimps of the deeper water zooplankton can discharge a luminous cloud into the water thus confusing, possibly even temporarily blinding an attacker. The emission of a luminous cloud by the deep-sea squid *Heteroteuthis dispar* has been likened to the defensive discharge of ink by shallow-water squid species.

In some organisms luminescence might be a lure to attract prey; this could be especially applicable to some fish.

Almost certainly, in many animals, the particular pattern of photophores can serve as identification for the species, possibly to enable a group to keep together, or even as a sex identification, since some species have sexual differences in the photophore arrangement. This would presumably be fairly limited since the distance over which the photophores would be discernible would be restricted.

An hypothesis has been put forward, supported by some experimental data, that some mid-water oceanic fish use their photophores for camouflage. The basis of the argument is that, in the lower depths at which light can be detected coming from above, a fish would appear as a dark shadow to another fish below it, thus making attack easier. If the fish has photophores and the distribution of light from these photophores is of the same intensity and quality as the descending daylight at the level of the fish, then, theoretically, it would be invisible when viewed from below. Certainly many mid-water organisms have downwardly directed photophores which, when coupled with the associated reflecting surfaces, would give just this effect.

No doubt luminescence has different functions in different organisms and probably all the above suggestions are correct in some degree. Equally probably there are some functions yet to be discovered.

7 Zooplankton as Food for Man

7.1 Nutritional value

Before any effort is expended in harvesting zooplankton for food, due consideration should be given as to whether or not it is worth catching.

The human body has the ability to synthesize certain essential compounds, but others, e.g. vitamins and certain amino acids, must be obtained from an external source. Should this not be possible disease or death will result. Proteins are essential for the maintenance of tissues and organs and, of course, for growth. They have a basis of some twenty amino acids. Some amino acids can be synthesized by the body; others, the so-called essential amino acids, must be obtained from the diet. A diet lacking any of these must be considered deficient. Zooplankton, particularly crustaceans, contains all these essential amino acids. On a dry-weight basis the protein value of the crustacean element of the zooplankton can be compared favourably with beefsteak.

The fat content of zooplankton can vary between about 5% and 10%. Some crustaceans lay down small reserves of fat during the cold months, having then a fat content of up to about 30%. Oil can be present in high proportions, particularly in the polar and deeper water populations; again it is the crustaceans which have most of the oil. The carbohydrate content of zooplankton can vary between about 12% and 20%.

Zooplankton, consisting to some extent of sea water, contains all the necessary minerals.

Vitamins are, to a large extent, an unknown quantity in the zooplankton as a whole. Some marine bacteria are capable of producing B_{12} as metabolites, as well as thiamin, biotin and nicotinic acid. These can be found in the surrounding sea water and it would not be unreasonable to assume that the zooplankton would contain at least some of the B vitamins.

Vitamin C is reported as being present in plankton, probably the phytoplankton. By contrast vitamin A or its precursor is particularly common in the eucarid crustacean element of the zooplankton. A biochemist has written '. . . the innumerable krill remain as a huge reservoir of concentrated vitamin A, on which all larger marine animals, and ultimately man, may draw'.

Thus there is every indication that, if not a complete food in itself, zooplankton would be a most valuable supplement to any diet, and is an especially good source of proteins which contain essential amino acids.

7.2 Suitability of different populations

In high latitudes there can be large zooplankton populations of either a single species or closely allied species, and it would be possible therefore to choose collecting gear to sample these populations. In the lower latitudes populations are much more mixed and there would be a large element of 'undesirables' in the sample.

A general division can be made between gelatinous and non-gelatinous animals. Although when expressed in terms of dry weight gelatinous animals have a high percentage of protein, the living animals are largely water (95% or more) so that there would not be a great deal of food value in a fresh sample. Additionally they have the reputation of being bitter and not at all palatable. Some of the tropical medusae have powerful, even lethal, stinging cells, and could represent a hazard unless the stinging cells are suitably de-activated.

This mixture of types is an argument against large-scale sampling for food in warmer waters; nevertheless as a life-saving exercise it would be of value. Sampling at night would be an improvement, since the proportion of gelatinous animals is usually lower due to the upward migration of various crustaceans.

The non-gelatinous animals are, effectively, crustaceans. Pelagic crustaceans are usually about 80%–85% water, have a high protein content and are much more palatable. In high latitudes individual species such as the copepods *Calanus finmarchicus* and *Paraeuchaeta norvegica* and the euphausiid *Euphausia superba* (the krill) can form virtually 100% of a zooplankton community.

The size of krill populations may be judged by an estimate, thought to be conservative, that in a square nautical mile there are some 140 million individuals distributed in patches. In addition these are relatively large individuals, being between about 15 and 65 millimetres long, the most common being between 20 and 30 millimetres.

These are accepted as more or less palatable—but there are disadvantages. A feature of all crustaceans is the exoskeleton, chitin, which is a polyacetyl glucosamine associated with protein. This can be very hard and, even in the smaller crustaceans, can be quite tough; it constitutes between about 5% and 20% of the dry weight of the animal. Some scientists are of the opinion that the ingestion of large amounts of chitin, particularly over an extended period, would cause physical damage to the gut, rather in the manner of powdered glass. The problem might not be serious in the case of copepods, but could be when dealing with krill. There are mechanical devices for 'shelling' the larger crustaceans, but this might not be practical with the smaller animals.

7.3 The practical aspects

Nobody can gainsay the importance of catching plankton as a life-

saving operation, but if such an operation is required to be economically
viable, then the situation becomes rather more complicated.

It is agreed that plankton would not be satisfactory as a complete diet,
but rather as a dietary supplement. The areas of the world where this high
protein dietary supplement would be most valuable are in the tropics; but
it is in the tropics that zooplankton individuals are generally small, there
are no aggregations of large crustaceans comparable with krill and,
although the annual plankton production probably exceeds that of cold
waters, the zooplankton does not occur in the concentrations found in
colder waters.

Analysis of local warm-water populations of zooplankton shows that
catching efforts should be directed towards copepods. A general figure
for the concentration of post-naupliar copepods in tropical shallow water
is about 1500/m³. After vertical migrations this figure could rise to
between 4000 and 5000/m³ in the surface layers at about dusk or dawn.
The size distribution will approximate to that shown in Fig. 7–1. If we take
the figure of 5000/m³ then the biomass would be about 168 mg/m³.

Energy sources in the tropics are limited, but labour is relatively cheap.
This suggests that nets mounted in shallow-water tidal streams worked by
men would be a more practical proposition then using powered boats
pulling nets through the water; the size of the net necessary would
require a large amount of power to tow it through the water. Let us
assume a tidal flow of about 2 knots. This is equivalent to a flow through
the net of about 1 metre/second. The net can be mounted on rotating

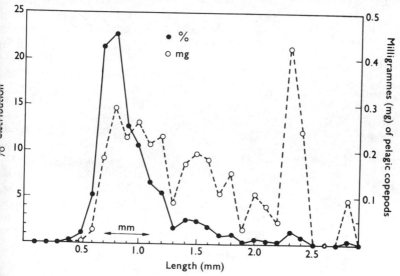

Fig. 7–1 Graph showing percentage length distribution and biomass
distribution per hundred copepods, in milligrammes, of pelagic copepods in
tropical coastal waters.

frames so that the mouths can be set to face the tidal flow. Suggested mesh aperture would be about 200 μm and the mouth aperture 3 × 3 metres. A flow of 1 metre/second means that about 9 m³ of water will be filtered every second. As the net will not be in the region of maximum copepod concentration a figure of 3000 copepods/m³ would be more realistic.

From this we can extract some figures. A mass of 3000 copepods/m³ represents a biomass of about 100 mg/m³. This is equivalent to a dry weight of about 17 mg/m³. The protein content is about 60% of this, say 10 mg. Every cubic metre of water filtered therefore represents about 10 mg of protein. Let us assume that the net fished for about 4 hours in every ingoing and outgoing tide—about 16 hours a day. The net would catch protein therefore at the rate of about 5184 g per day. Daily protein requirements for humans are about 1 g/kg of body weight for adults and about 4–5 g/kg for young children. Adults thus need about 70 g of protein daily. It looks, then, as if the net can supply the daily protein requirements for about seventy-five adults.

Clearly there would be many snags. The water would no doubt be thick with phytoplankton at different times and there would be a good admixture of non-copepod and non-crustacean animals so that some means of separation would be necessary. However, although a doubtful commercially economic proposition, where suitable conditions exist, it is a means of supplementing a protein-deficient diet which should be given serious consideration.

As a modern commercial enterprise the only operation which can be considered is catching the huge numbers of krill in the southern oceans. There have been euphausiid fisheries for many years off the coast of Japan, off Norway and Western Canada, and in several other areas, but these are all on a small scale; only a small part is destined for human consumption.

For the last ten years or so the U.S.S.R. has put a great deal of effort into investigating suitable methods of exploiting the huge resources of the antarctic krill, *Euphausia superba* (Fig. 7–2). Other countries such as Japan, Norway and the U.K. have also undertaken investigations, but on a much smaller scale. An analysis of this species has shown the following: 81.6% moisture, 10.3% crude protein, 5.7% hot-water-soluble protein, 2.0% carbohydrate, 3.4% crude fat, 2.7% ash and A and B vitamins at fairly high levels—all in all, a very nutritious animal.

One estimate considers that it would be possible to catch about 70–100 million tonnes of krill per year without depleting the stocks. This represents a greater weight than the total fish yield.

This involves processing of the krill on a very large scale, and the krill are washed and pressed in a continuous press. Then it is given 10 minutes heat treatment to coagulate the protein, which is then separated, frozen and stored in blocks at about −20°C.

The produce, although undoubtedly nutritious, is not entirely

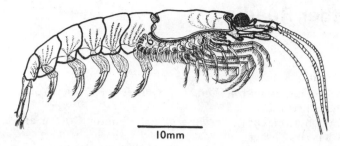

10mm

Fig. 7–2 *Euphausia superba*, the antarctic krill.

acceptable aesthetically as human food, and much of it is used as animal feed. A patented procedure to enhance its acceptability is in use.

The operations involved are highly mechanized and demand relatively sophisticated equipment. It would seem that because of this, and the many processes involved, it is not yet generally accepted that catching krill, even on this scale, is economically viable. However, increased efficiency and better processing, plus world pressures for highly nutritious foods, may yet result in zooplankton fisheries in general, and krill fisheries in particular, being a commercial development of the near future.

A point which merits further consideration is at what stage of the food chain should organisms be withdrawn to feed man. Formerly a figure of 10% was taken as being reasonable for the energy transfer from one step of the food chain to the next. This means that for, say, a fish to put on 10 g in weight it would need to eat about 100 g of zooplankton. Experiments now show that transfers of energy can be much higher than this, up to 40% or even 60%. We should accept that animals high in the food chain can be more efficient than man in capturing and utilizing the food below it in the food chain.

Consider that a single fish about 35 centimetres long will have a protein weight of about 70 g, this single fish thus providing the total daily protein requirements, in quality and quantity, for one adult. Might it not be a better proposition to catch a single 35 centimetre fish than about 225 000 copepods?

In the case of the krill, would it not be a more efficient utilization of their resources to have a controlled programme of breeding and capturing baleen whales than to attempt to harvest the krill?

As a final point, mention should be made of the possibilities of suitable zooplankton animals being cultured. This has been quite successful in the case of one or two species, but on a small scale. Cultural purposes are usually either to provide experimental animals or to provide suitable food for rearing the early stages of fish species. With regard to culture in order to provide human food, the scale required would be quite enormous, and is very much a future proposition.

Further Reading

CLARKE, M. R. and HERRING, P. J. (ed.) (1971). *Deep Oceans*. New York, Praeger Publishers Inc.

COSTLOW, J. D., (ed.) (1971). *Fertility of the Sea*, London, Gordon and Breach Science Publishers.

FRASER, J. H. (1962). *Nature Adrift*. London, Foulis.

HARDY, A. C. (1956). *The Open Sea, its Natural History: the world of plankton*. London, Collins.

HARVEY, H. W. (1955). *The Chemistry and Fertility of Sea Waters*. London, Cambridge University Press.

LEITH, E. N. and UPATNIEKS, J. (1965). *Scientific American*, 212 (6), 24–35.

LINKLATER, E. (1972). *The Voyage of the Challenger*. London, John Murray.

MARSHALL, N. B. (1954). *Aspects of Deep Sea Biology*. London, Hutchinson.

PICKARD, G. L. (1964). *Descriptive Physical Oceanography*. London, Pergammon.

RAYMONT, J. E. G. (1963). *Plankton and Productivity in the Oceans*. London, Pergammon.

RILEY, J. P. and CHESTER, R. (1971). *Introduction to Marine Chemistry*. London, Academic Press.

RUSSELL, F. S. and YONGE, C. M. (1963). *The Seas*, new and revised edition, London, Frederick Warne.

SVERDRUP, H. U., JOHNSON, M. W. and FLEMING, R. H. (1942). *The Oceans, their Physics, Chemistry, and General Biology*. New York, Prentice-Hall Inc.

WIMPENNY, R. S. (1966). *The Plankton of the Sea*. London, Faber & Faber.